Trains Don't Fly

J.F. Frana

First published 04/2010

Printed in the United States of America

Dedication

I wish to dedicate this book to the three geniuses that it was my privilege to work for. They not only taught me much about the railroad, but about life in general. I am also dedicating it to all the people I worked with. It is as much their story as mine.

Acknowledgements

Coleen, without her this book could not have gone to print in its present form.
Renae, for all her organizing and photo work.
The city of Cresco, Ia. The pictures of the railroad equipment were taken in the city park.
Calmar, Ia. Historical Society, which is where I got the picture of the train order hoop.
My wife, for her good nature as I wrote this.

Cover Photos:

Engine on the front cover is a F7 manufactured by General Motors in about 1950. It was commonly called a covered wagon or a Cadillac.

Back Cover is a caboose from the same era.

Definition of Job Titles

Agent - Represented the railroad in a town they served. He may also have acted as the operator, express man, and handled telegrams. It was His responsibility to collect freight charges and do whatever other Business the railroad may have had there.

Brakeman - Main duty is to handle switches. Also assists the engineer in Watching for signals. Sets handbrakes on cars the train sets out.

Carmen - Repaired railroad cars, re railed cars and engines

Caller - Called train crews to duty

Chief dispatcher - In charge of day to day railroad operations. Hired agents, Operators, and dispatchers, and made sure these positions were staffed

Conductor - Man in charge of the train. Does the bookwork and handles the Documents (waybills) that accompany the cars the train is hauling.

Deadhead - Travel, such as from a home terminal to an outlaying point to get a train

Dispatcher - Responsible for the movement of trains. Chief duty was to keep them from colliding.

Engineer - Runs the engine on the train

Fireman - Tended the fire on a steam engine or assisted the engineer on a diesel. Usually became an engineer.

Lead - A track in a yard leading from one point in a yard to another, such as from the roundhouse to the main track, or a track providing access to other tracks

Operator - Copied train orders, sent and received messages on the telegraph

Road master - man in charge of track maintenance on a division of the railroad

Section man - track laborer

Section foreman - the man in charge of track laborers

Signal maintainer - Maintained both signals governing the movement of trains and those at road crossings

Superintendent - In charge of a division of the railroad, supervisor of all who worked on that division often had an assistant who may also have assumed the duties of a trainmaster at an outlaying point.

Switchman - Switched cars in a railroad yard, making up trains, placed cars for loading or unloading

Trainmaster - Supervisor of the train crews, conducted signal tests, hired brakeman and switchmen

Traveling engineer - Hired, taught, and supervised the engineers. May also have assumed the duties of a trainmaster

Yardmaster - In charge or a railroad yard, chief duty is to supervise the Making up of trains.

FORWARD

This book contains a collection of stories, events, and circumstances observed, described to me and experienced by yours truly over a 42-year railroad carrier, spent mostly as a train dispatcher. A train dispatcher is to the railroad industry roughly what an air traffic comptroller is to the airline industry.

During the Regan era strike by the comptroller, some of these people became train dispatchers when they lost their jobs. I was recounting to a friend of mine my experience with an air traffic comptroller that I was teaching. After hearing the story, he suggested that I write a book about my experiences.

My student was engrossed in another project when an engineer called in on the radio. I knew what he wanted. His train was moving on what is called track warrant authority and he was about to reach the end of its limits.

The content of the next warrant issued to his train would determine where he and an opposing train were to meet. I pointed to the two trains on the sheet where train movements were recorded. The student nodded calmly and in a very business like manner, instructed the engineer to increase his altitude by two thousand feet. Neither the engineer nor I thought these instructions to be viable. This is where the name of this book came from. The human mind is a strange complex thing. It tends to do what it has been trained to do.

I will attempt to not bore the reader with any more railroad rules and regulations than I have to. At times some background information will be necessary to give events meaning to anyone not familiar with railroad procedures. Also, humor, like beauty, can sometimes be in the eye of the beholder. Many of the subjects visited will be of a humorous nature, at least to some. I do not expect to entertain all of the people all the time, but would very much like to entertain some of them at least some of the time.

The passage of time tends to dull a persons memory. There sometimes were more than one version of the events I heard about.
I have tried to give the best account I could under the circumstances. The sources I quoted were probably reliable, but they did not have their hand on a bible.

There is a glossary in the back to help explain the terms used throughout the book.

CHAPTER ONE THE SCHOOL HOUSE

Farm Boy Goes To Railroad School

Just out of high school and fresh off of an Iowa farm, I found myself in the big city attending my first class at the railroad school. It was a class on railroad rules. The instructor, who I later found out, was a retired superintendent from the Great Northern Railroad, held three one hour classes a day. Dividing the students in thirds kept the size of the classes down to a manageable number. He gave a short talk and asked if there were any questions. One fellow raised his hand, and asked how well he wanted us to learn these rules. The instructor was holding a railroad rule book, some 250 pages long, and he threw it at the student.

"Open it," he told him very gruffly. "Now read the first three words of the first full paragraph on that page."

The student did as he was told. The instructor then told him what page he was on and quoted the rest of the page, word for word. "That is how well I want you to learn your rules. Now bring my damn book back."

This rule book had pages that were somewhat smaller than a regular book, but it was enough to impress me. I also got the impression that this man was not somebody that you wanted to get smart with. PR was not his long or strong suit.

The next class, with my same classmates, was in railroad accounting. This class was taught by an auditor who had retired rather young on disability due to some health problem. His style of teaching was the exact opposite of what I had encountered in the previous class. He tried to bring a little humor into whatever he was trying to teach us. He was an excellent teacher and was extremely well liked.

Then came the telegraph courses. Lots of Morse code. Three words a minute to start. It did not take me a great deal of time to discover that I was not a natural-born telegraph operator. I sat in this class until there was only a hand full of us left. One day, a light came on. I managed to pass the test and move onto six words a minute. When I arrived there the instructor told me that he was about to hold a test. I should sit down and try to copy what I could.

To my amazement I passed the test with a perfect score. Ten words a minute was the next step up. This class was held in the other end of the same large room as the three words a minute class. Unlike the first class, where you got some personal attention, here they just ran tapes. Practice, practice, practice.

The instructor, who was in charge of both classes, was more than somewhat surprised to see me. He asked if they had sent me back because I was unable to copy anything. I told him that I had passed the test. He just smiled, and told me that this happens once in a while.

When I first sat down at a ten words a minute desk, I was not too sure that passing the last test was such a good idea. Everything was a blur to me, not being able to pick up even a single character. I sat there trying until the end of the period. I told the instructor of my dilemma and asked his advice. "You are trying to copy the whole works at once," He told me. "Try to pick up only one letter, e to start with, at a time. Be patient. You are now ahead of half the class." This eventually worked, but I never became any hotshot telegrapher. I finished in about the middle of my class. I did not feel this was too bad. There were quite a number of former radio Morse operators going to school with me under the GI bill. They were able to make the transition from one code to the other quite easily.

The Real World

I was to find out in later years that the telegraph one heard while working was quite different from what one heard at the school. It was not a stupid tape, but a human being that one heard. They all had a different style. After a while, one could tell who was sending just by his style.

When I started dispatching, we had an operator in a town that was a crew change point. He sounded so different when he saw the train coming that one knew it before he told you. When he called you to ask something, his sending sounded relaxed and almost lazy. When a headlight was showing, he meant business. He wanted to get something done.

We had one fellow that I always maintained broadcasted, not sent. His Morse was always the same speed, every character perfect, spaced properly, almost like everything was in capital letters. He was the first guy that I could copy easily. At the other extreme we had this old agent who was the exact opposite. He used one side of his telegraph bug to make both his dots and dashes, the dot side not functioning. He sent pure garbage.

It took me four or five years to learn how to copy him at all well. He could send decent Morse if he wanted to, he was just too ornery to do so. He thought being he could copy this kind of crap, everybody could.

I had been working several years when a relay operator was in the office visiting. He maybe copied code for six or seven hours a day. The old agent called me and gave me a list of cars the way freight had for me that day. I said "ok" after I had copied the information.

The relay operator asked "What the hell was that? The only thing that I understood was your ok." I guess I was not the only one to have problems understanding him. I was working a job that required me to send him a long train list every day. By this time I could send rather well and adjusted my bug to the fastest speed that I could handle. I would pour the Morse at him so fast that my hand and wrist hurt by the time I was through. All I ever got back from him was a very sloppy "OK." Never once did he ask me to repeat myself. It maybe was darn good Morse, but I could not have even come close to copying it at half that speed.

It was 1953, and there did not seem to be a shortage of jobs anyplace. As soon as the chief dispatcher in the office where I first applied found out that I could telegraph a little he hired me. I lacked being eighteen years old by about a month. He informed me that I would have to wait until my birthday before I could go to work. That was the law. This suited me fine. This railroad serviced the area where I grew up. I felt I could use a little time off after months of both going to school and working full or part time.

Telegraph bug

CHAPTER TWO THE EXTRA BOARD

My First Train Order

A couple of weeks went by, and one day the telephone rang. It was the Chief Dispatcher that had hired me. "Got a couple weeks work for you, starting Monday."

My reply was "I'm sure glad to hear that, sir, but I am afraid that I still won't be 18 by then."

"Oh, shit. Well, I'll call you next time something comes up. Thanks." And the connection went dead. Hearing from him made me feel good. At least I knew that he had not forgotten about me. I hoped I had at least made half of a brownie point with him by pointing out my age.

In later years, I was to discover that the chief dispatcher was a very busy man. This one had to worry about staffing the agents, operators, and train dispatcher's positions over about two thousand miles of railroad. That was only a tiny portion of his various duties.

True to his word, the man called me sometime later. I now held a job at a tiny town for two weeks. People on the extra board worked wherever someone was on vacation, sick, or otherwise absent for some period of time. The man knew where to send me. Not much went on there, and it was an ideal place for a new man to learn the ropes.

Most stations were very much alike in where things were kept. The routine in each was much the same. Most of the neighboring agents were very kind in giving a new man advice. One had the chance to meet many new people in each place one worked. I was to work the extra board for about two years, at which point I had enough seniority to hold a regular job.

I never will forget the experience of handing up my first train order to a train. I had maybe all of three days experience when my dispatcher's phone rang, late in the afternoon. I had

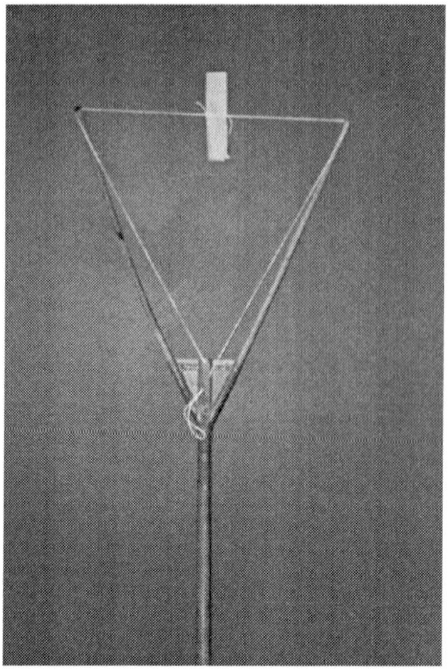

no trouble copying the order. It read "No. 79 meet No. 94 at" and gave the name of the station. It was dark. It was snowing and blowing. No. 79 was a passenger train, maybe four or five cars long.

I placed the orders, tied with string, in the hoops, which are little more than forked sticks. I managed to light a kerosene lantern and set it on the station platform, a little way down the track from me. This gave the train crew some light. I did not have to wait long before the headlight showed up. At least I did not have time to worry much about what I was about to attempt.

I had been told that the exact positioning of the sticks was not as important as keeping them still. The trainman catching the orders could adjust to the position. But if you move, it would be hard for him to grab them. I stood as close to the track as I dared I stuck one stick up in the air far enough so that I thought the engineer could reach it, and the other out about letter high. The headlight looked huge; I stood as still as I could. All of a sudden I felt two jerks at almost the same time, and the train was gone.

At first I could see nothing but blowing snow. The lantern was all wet from the snow. I could see that the strings and orders were both gone. The train was going about seventy miles an hour. It amazed me how much snow it had thrown up. I had snow in my hair, my ears, and who knows where else.

I reported the train by to the dispatcher and he asked me if they got the orders ok. I imaged he, knowing that I was green as grass, had been concerned. He did not want to see the train have to back up to get them.

These were the most adverse conditions I ever encountered handing up a set of orders, so it got easier after that. I later found that the train crews were very skilled in picking up orders on the fly. They had a lot of practice.

The station agent of that day worked for three different companies, the railroad, Railway Express and Western Union. The railroad paid a salary, the other two, a commission. Even though telegrams were sent to and received from most any spot on earth, this paid very little. I have copied them in Norwegian, Czech, and German, not my favorite way to make a few cents. Express was something else. It could be very lucrative at some stations. The express company treated the agent very well.

The only negative experience I had handling express was when I worked an operator's job and handled cans of cream. About nine-thirty each evening several farmers came to the depot and I would wind up with about thirty cans. The cream moved to a creamery that was not very far away. The express charge was nine cents per can. We handled the empties that came back for nothing. The commission was 10 percent. That was the good part. The bad part was I had to split the commission with the agent.

Egg On My Face

My first claim to fame was my ability to lose a railroad car on a track that was maybe less than a quarter of a mile long. Well, maybe notoriety might be a better word than fame.

I came to work one morning and noted an open top coal car that was being unloaded when I left work the previous evening was no longer there. I did not see the car anyplace. I assumed that the crew on the night train, finding the car empty, had picked it up. I reported the car as such.

That afternoon I was out collecting freight bills when I saw the car on the extreme west end of the track. It was parked behind an old storage shed. The curvature of the track and the building obscured it unless one got very close to it. With a lot of egg on my face, I reported my error.

An experienced man would have known it was common practice to leave such a car where it was very handy for the next crew to pick up. He would have looked there at once. The chief dispatcher did not even bother chewing me out. I imagine that he thought that I felt sheepish enough already. But I was afraid that my half of a brownie point had vanished. I got a lot of good natured ribbing about this incident for the next twenty years or so. At which point, I think most of the people that knew about it had retired.

All in all, I would say that I have little but fond memories of this period. The only bad situation, if one could even call it that, was when I was sent to a station where the agent had been injured in an automobile accident on a Wednesday evening. No one had been working the station until I got there the following Monday morning.

Every industry track in town was full of cars. The freight house was full of goods for the local merchants and the express room was almost as bad. The Western Union operator was calling me on the telegraph wire to copy delayed telegrams. The work load at this station was more than I was used to on a good day. This was not a good day, to say the least.

I managed to get the car numbers of the cars on hand by wading through the snow banks. I was the last agent to send in the telegraph report on the status of these cars that morning. I still do not know if I got it right. A guy showed up wanting to ship a barrel of frozen tame rabbit to Chicago. There had been a drought in Oklahoma. Two parties were shipping hay there. Both wanted to know when they would get some service. Then someone wanted to buy a ticket on the evening passenger train to someplace in Florida.

The sweat was starting to roll by the time the traveling auditor arrived about ten o'clock. It was after eight that evening before we caught up. The auditor not only did a very big half of the work involved, but took time to teach me. I maybe learned more in that one day than I would normally learn in a month. I am grateful to him to this day. I did not put in for any overtime. I hoped I got my half of a brownie point back.

When the first seniority list that had my name on it came out, I was number seventy four. The last one. The guy at the top of the list had a date of June 14, 1898. He had only fifty-five more years of seniority than I had. It looked like it would be a long time indeed before I made the top of that list.

It was about five one evening when the division lineman asked me if I wanted to take a ride with him. He had a problem with the telegraph wire. The little motor car he used was designed for two men and a few tools. One man could set off at a crossing, but not anyplace where the wheels would sink into loose rock or dirt. We had a freight train ahead of us, and would have a passenger train behind us. He wanted me along in case he had to set the motor car off where the trouble was.

We took off, a spotlight pointed at the line. It was dark, with snow on the ground. It thawed that day and the warm air was creating some fog. We went a few miles before we spotted the trouble, a broken tree limb in the wires. By the time we got the limb down we did not have enough time to run back in the direction the passenger train was coming from. There was a branch line three or four miles away. We headed for it.

The motor car did not do too well up the hill. We were too big, the motor too small. We got over the top of the hill and the lineman wanted to make up for lost time. We were going along at a very good clip when I became familiar with a device called a torpedo. It was sometimes also referred to as a gun.

We were almost to the bottom of the hill when we were jolted by an explosion. It sounded like a cannon fired at us at point-blank range. A ball of fire larger than a basketball accompanied the blast. It felt like the motor car had been raised in the air. After we checked the family jewelry, we dug out a red fusee and lit it. This was to protect both ourselves and the train that maybe was stopped ahead of us. A good mile later, we arrived at the switch and got in the clear on the branch line, tossing the fusee into a snow bank. Soon the passenger train went by.

A torpedo was used by a flagman to protect the rear of his train when it stopped and could be overtaken by another train. It was a flat packet of explosives attached to two soft metal strips that could be molded around the rail. The following train, hearing the explosion, had to slow down to twenty miles per hour for the distance of at least one mile.

The use of two torpedoes was required but the explosion of one meant the same as two.

Evidently, the train ahead of us had trouble and had to stop. His flagman had been busy.

I do not know what happened to the second torpedo. Maybe our motor car was too light and did not set it off, which was good. In all my years on the railroad, this was the only time I ever heard a torpedo explode. And that was one too many. It scared the daylights out of me.

A Trip To Fantasy Land

Shortly after starting work at one station I got a call from what sounded like a sweet old lady wanting to make a trip by train. She informed me that she always tried to see the desert when it was in bloom. I wrote down a rather lengthy list of what she wanted, and had the information ready for her when she called back. She did not want to be on the train all night. I had to figure out where she might stop and how her ticket had to read. She was going to be gone about ten days.

She wrote all of the information down, asking for the names of some of the places the various trains went through. When I asked her if I should get her reservations booked and her ticket ready, she said no. she said she could buy her ticket at the depot she was leaving from, she never had a problem with getting a seat, and the railroads were not that busy anymore. I thought this to be rather odd, but it was her trip.

Two weeks or so later, she called again. This time wanting to go through the Shenandoah Valley, which she said was very nice this time of year. We went through the same process again. When I asked her how she had enjoyed her last trip she proceeded to tell me about all the beautiful blooms she had seen. She thanked me profusely for arranging such a nice trip for her.

When I worked the station the following year, she called again. Another trip was laid out for her. I happened to run into the regular agent one day. Curious to find out if this lady traveled all the time, or just happened to do so when I was there, I asked him about it. He laughed and told me this woman was housebound; she traveled only in her imagination and had been doing so for several years. He also told me he had given her many imaginary stations and fares. Well, at least that way she had no trouble getting a seat, as I imagine that she was already sitting in her rocker. I guess this was a rather harmless way to amuse oneself, and that I would get old someday, too.

A Weighty Problem

Not too long after this I acquired enough seniority to hold a permanent job. As luck would have it, the job was close to where I was reared. I also soon found out there was a difference in holding a regular agent's job. I was no longer that stupid kid off the extra board, but a regular man who was expected to know the answers. I guess nothing is all good or all bad. I did not feel that I had anything to complain about.

I was by now a half way decent telegrapher and prided myself on my sending. I could understand most everything said on the wire if I had my feet up on the desk, just listening. If I tried to copy real fast Morse down long hand, I could not write fast enough, or could not read what I had written. I never could get the hang of using a typewriter very well. The noise from it interfered with the sound of the telegraph. But I was good enough to get by very nicely. I sent very good Morse to the operator that I happened to be working with at a moderate rate of speed; he would do the same for me.

As a train dispatcher, I liked working on the wire very well. A train order was transmitted, and copied in the book the first time an operator repeated it. Then it was underscored with each repetition. There would be a bunch of old head agents on the line that could hear what was being said wherever in the office they happened to be. If someone made a mistake, it would get caught at once. This was a nice safety net. Most anyone would come in on a dispatcher's telephone and bother you. They had to be telegraphers to use the wire, so you had peace and quiet in this respect. This was maybe my favorite way to dispatch trains using train orders.

One day the crew on the local, which stopped at about every station, sent word to me that I should have some help on hand. They had a very heavy safe to unload. It was a very nice day and there was nobody in town. The farmers were all in the field. As close as I could find to any help were two old drunks sitting in the tavern, one of which had a hard time walking even when he was sober, which was rare.

The train pulled into town and I told the crew that I had not been able to find anybody to help. We opened the box car door and backed the heavy baggage cart up to it. There sat the safe, over a ton of it, smiling at us. The crew took one look at it and decided to all go up town to look for help. As it was a warm day, they had an excuse to get away for a quick beer or two.

Shortly after they left, a farmer who lived across the track pulled up. We had his private crossing blocked. He wondered how long the train would be there. I told him what the situation was. He looked the safe and cart over, and asked if there was any pipe around. There were a few short pipes in the freight house. "Come on," he told me, "I will show you how to get that thing out of there."

I had used a big pry bar to knock the blocking that was around the safe to keep it from shifting loose. We used it as a lever to lift the safe enough to get a pipe under the skids that it sat on. Then we did the same on the other end. Now, with the pipe rolling under it, the safe was very easy to move. We moved it far enough so that it balanced on the lead pipe, and then repeated the process. Inside of five minutes the safe settled on the baggage cart with a resounding thump, the cart being slightly lower than the box car floor.

The crew was not back yet. The farmer decided to drive around the train, park on the street, and walk the short distance to his house. I thanked him, and he drove away smiling. He was the typical farmer of the era, very self-reliant and even more willing to help anyone in need.

In due time the train crew came back. They were more than a little surprised to see the safe sitting on the cart. I informed them that I could not spend all day waiting for them and told them that the least they could do was to help me wheel the cart into the freight house. I never did tell them how that safe got on the cart. They all went to their graves thinking that I had unloaded it alone. Served them right for drinking beer on duty, or so I thought.

The safe is still in service. I guess we did not damage it when we moved it. It outlived the building it was first moved into. It looks heavier now than it did fifty some years ago. The postmaster was kind enough to take a picture of it for me. He did complain that the tumblers now stick once in a while, and he has to give it a whack to get them to drop. I told him if he has to ever replace it, about all he can expect from me is a little moral support when he moves it. Once was enough.

Getting All Steamed Up

We still had a steam engine on the way freight, one of the last trains we ever ran with steam power. One day the way freight had to meet another train at my station. He about stalled pulling into the siding, barely able to get into the clear. The engineer and fireman were cursing each other, the engineer all upset because he was out of steam. It turned out there had been a derailment. A car of coal was dumped, salvaged, and used as fuel for the engine. It was some sort of smokeless coal. The fireman, watching the smoke, had overfed the firebox. Coal spilled out when they opened it. They ended up going to the hardware store and buying a propane torch to get the fire going. They were now laughing about the situation instead of cursing.

We had another fireman that used corn cobs to get a hot fire going. I remember the old agent telling the dispatcher, "Well, the corn cobs just went up, the way freight should get out a little early today." He was right. The fireman walked by the depot with half a grain sack of them, and could get a head of steam up in an engine a good twenty minutes sooner than anyone else. I know as a kid I have seen wood burning cook stoves ruined by using too many corn cobs. I guess they must throw off a lot of heat in a hurry.

There was much more involved in working a fireman's job than shoveling coal. By this point in time most steam engines had a stoker in them. A conveyer moved the coal from the tender into the firebox. The coal fell on to a large dished plate and then distributed by jets of compressed air. The jets were adjustable, allowing the fireman to keep his fire level, which was very important. If a fire was not kept level it would not produce enough heat to keep the steam pressure high enough and allow huge clinkers to form.

The ideal pressure for the boiler to be kept at was two hundred pounds per square inch. This was known as a full head of steam. Less than this, the engine lost power. Too much more, there was the danger of the boiler exploding. There was a sight glass that told the fireman how much water was in the boiler. This had to be monitored closely, as a shortage of water could also cause the boiler to explode. An experienced fireman could tell how his fire was doing by the appearance of the smoke coming out of the smoke stack, he watched it closely

The quality of the coal the railroad used many times left

much to be desired, it contained many impurities. Even the most skilled of fireman could end up with clinkers in his fire. They would get almost as big as a wash tub. The train would have to stop to clean the fire. A huge crow bar was used to break up the clinkers before they could be removed. This was not a pleasant job, especially on a hot summer day.

In flat country a large steam engine could handle about four thousand tons quite well. The top speed for many steam engines was about a hundred miles an hour. The driving wheels of a large engine stood about seven feet tall. If more than one engine was necessary to pull the train it required a second engineer and fireman. They could not be placed in a line, one engineer controlling all the engines, like diesel power can be. I think that this fact played a big part in the diesel replacing the steam engine. There were also steam engines that burned fuel oil, but I do not believe I ever saw one of them.

CHAPTER THREE THE TRANSITION

Again A Student

During this period of time the railroads were both heavily taxed and regulated. This is a separate issue, so I will try and stay off of my soap box, but I could see the handwriting on the wall. The railroad had to trim expenses some place, and started by closing small stations and abandoning branch lines. The seniority I had acquired over the past years was starting to evaporate. By this time I had a family to think about, so I decided I would try to become a train dispatcher.

I managed to acquire a station that was very close to the dispatcher's office. I was familiar with a good part of the railroad involved and had experience operating a centralized traffic control machine. Ten years experience as an operator was sort of the minimum required. This I had, barely. I asked the chief dispatcher if there was any chance of getting a dispatcher's job. "Well, kid," he told me, "hang around the office for a while on your own, and if things go well, I will see what I can do."

Hang around the dispatcher's office I did. Evenings, weekends, whatever. For the most part, I got a warm welcome and lots of assistance from the men working there. Time went by, and the chief called me aside. "Kid," he told me, "I can get you a thirty-day rate sheet and can start paying you. It will take a good six weeks to get it and by that time you should be ready to really start learning. Just remember that you will have a rules test to pass before you can go to work on your own, so start studying."

Study I did, enlisting my wife, one of the dispatchers, or anyone else I could find to help me. I was familiar enough with the book of rules so far as an agent operator was concerned, but back in those days trains moved under a very complex set of rules. I had to learn all of them. Several long books, with titles such as "Rights of Trains," and "The Superiority of Trains," were written. I tried reading one of them. My knowledge of the rules was not good enough to grasp much of what they covered. This book of rules itself had been written mostly by disaster, over a long period of time. Many of these rules were in the book to try to keep the same thing from happening again.

I concentrated on the rules that I was not familiar with. By the time I was through I did not know what some of them meant, or how to apply them, but I did manage to memorize them. I was starting to feel enough like a parrot that I sometimes had an urge to examine myself for pin feathers.

Time went by, the rate sheet arrived and my thirty days flew by. It was time to take my rules test. I passed. I was presented with a card, duly signed by the rules examiner, stating that I was now a qualified train dispatcher. My promotion was official.

There are many jokes about train dispatchers. There is one about the new trainman that worked almost a year before he found out that Goddamndispatcher was more than one word.

There is another about the trainman that died and went to heaven. While waiting for St. Peter to check him in, he noticed a man walk by, he was wearing a green eye shade, sleeve protectors, had a train sheet under one arm and a pen in the other hand. He asked St. Peter if that was a train dispatcher. St. Peter replied "No, that is God; sometimes he thinks he is a train dispatcher."

I guess that so far as train movement over a railroad is concerned, the train dispatcher was rather like a God. His word was final. I am sure that this attitude sometimes carried over into our personal lives

I can only state when the big day arrived and I went to work on my own, I felt more like a scared pile of dog meat than God-like. In those days there was little more than the seat of the train dispatchers pants and his pen to keep the trains from running into each other. The responsibility was rather crushing for a new man. It made one watchful.

There is a saying that it is very lonely at the top. I would imagine that a train dispatchers chair is nowhere near the top, but for me it was high enough to feel very lonely indeed. When I had finished my shift without mishap I felt rather like Superman until I realized that I would have to do the same thing again tomorrow. That kept me humble.

Impatience Can Be A Virtue

As I got to know the men I now worked with I was amazed at how different they all were. The oldest man in the office worked what was known as the west end, 8am until 4pm, with weekends off. I had spent the least amount of time breaking in with him, not only because I thought I would never get to work his job, but because he must have been one of the most impatient men on earth.

As an example, he did most of the grocery shopping, his wife taught school. If there were not enough check out lanes open to suit him when he finished shopping, he thought nothing of filling his cart to the top with ice cream, parking it in some obscure spot and doing his shopping elsewhere. The help there knew him and tried to watch him. They did not always succeed. His energy was almost boundless. His lack of patience did not seem to impede his ability as a train dispatcher.

I had only a few days experience when I was called upon to work his job. The man had unexpectedly needed a week off. All the other men in the office did not want to change their days off because of whatever they had going on. The first day I went to work I was sweating bullets. Not only was it the end of the railroad that I was the least familiar with, but I had spent hardly any time at all on this job. I soon found out the man had his operators so well trained they would come in without being called, saying it was time to do this or that, and this or that train needed something. I went home feeling more like a well paid stenographer than a train dispatcher. I never had an easier week of dispatching than that one. Maybe impatience does pay.

At some point in time I found out that several officials, all formerly dispatchers, were high ranking officers in the Army Reserves. They had all served as train dispatchers in a railroad battalion during World War Two. Our army ran many of the railroads in Europe toward the end of the war and after it was over.

A conductor in a railroad battalion had to be at least a sergeant. The train dispatcher had to out rank him. Any dispatcher who wanted to dispatch trains in the army was assured the rank of Lieutenant when he got out of basic training. Some of these guys did not have a great deal of patience either.

The Cool Cucumber

The man who worked first shift on the East end was completely different. He expended a whole lot less energy and was a genius at finding an easier way to do most anything. He would establish a meeting point for two trains when he found out when the crews were called for duty. As the day progressed, he would add or subtract a stop or two as needed. This was much easier than issuing a new set of train orders to both trains to change the meeting point and worked just as well. With him, easier usually meant better.

As an operator I discovered that sometime between noon and one there was always a period of about twenty minutes that he was busy on the telegraph wire. One would get a blast of Morse code in his ear if one called him on the dispatcher's phone. I came to find out during this period he sat there with one hand on his telegraph bug, eating his lunch and reading a bridge magazine. This was the most important task of the day, not to be interrupted. He was not talking to anyone. The blast of Morse was to chase you off the phone.

Over half of the territory we dispatched had no dispatcher's phone, only a telegraph wire. He worked the wire at a very moderate pace, nobody seemed to ask him to repeat anything, nor did he ever have to do anything twice. He seemed to accomplish as much or more than someone who did things hurriedly.

One of his duties was to keep up a gross ton mile report, which was meaningless except as a guideline of revenue expected. The chief noted, for three years in a row, we made over four hundred unexplained miles over a segment of railroad that was sixty-nine miles long. He told his clerk he did not give a damn about the miles, but they were going to find an explanation for them. No mater how long it took, he was curious.

It turned out the dispatcher had used seventy miles because it was easier to work with. Working some two hundred days a year, doing the report on two trains a day came out to the extra miles involved. When the chief jumped the dispatcher about it he laughed, commented that they must be used to the seventy-mile figure by now, and used it for as long as he worked the job. I came to the conclusion that whatever this man lacked in ambition was more than made up for by his intelligence.

I think this dispatcher came to worked dressed in a white shirt, suit and a tie every day of his life. He was a heavy smoker, and most of his pockets contained stick matches. When I had first started to learn train dispatching, he told the chief there was a bridge tournament that he wanted to go to that Sunday. He asked if it was OK to leave me alone to run the centralized traffic control machine which had somewhat recently been moved into the dispatches office. I had maybe run this machine as an operator more than he had. As there was little else to do on a normal Sunday, I had no problem saying yes. The chief told him that if he was comfortable with me running it, he could give a bleep less.

He got me started Sunday morning and left before ten. He lit a smoke as he was leaving, tossing the match into the wastebasket. The man working the chief's job followed him out the door on his way to church. They no more than left and the contents of the wastebasket burst into flame. I managed to get the thing into the restroom and get the fire out. There was no smoke left in my office by the time the chief got back. He stopped to use the restroom and excitedly shouted to me to call the fire department; the men's room was burning. Nothing like having the place catch fire the first time that you are entrusted with it.

Looking back, I think that the man wanted me to experience being there alone as much as he wanted to play bridge. It gave me some much needed self confidence, and was maybe the best thing he could have done for me. He also eventually taught me to play bridge. I became very fond of him. He taught me much, and I think had more to do with making me into the dispatcher that as I was to become than anyone. I owe him a debit of gratitude I can never repay.

All Work And No Play Makes Jack A Dull Boy

The man working third shift was probably closest to the typical train dispatcher. He did his work well, did not put up with much nonsense, and went home. However, he was not above having a little fun when he had the chance.

We had a petty official almost everyone seemed to have a problem with. Part of our duties was to maintain what was called a tie-up sheet. Starting with a fresh preprinted sheet each day, it listed the division officials. There was a box to check if they reported being home, a box for the name of wherever they were staying, and another for the phone number there. The superintendent we had was a real stickler when it came to expense accounts, often checking the account against the tie-up sheet.

I think this official liked to tipple a little on occasion and would neglect calling in so he would not be forced to go to work under the influence. The superintendent threatened to cut his expense account if he did not call in. In an attempt to get around this little problem he instructed the third shift man to show him tied up SOMEPLACE when he did not call in. The dispatcher told him he did not want to falsify the records, but the official became insistent and abusive.

One morning when I was breaking in, the superintendent came into the office, shoved an old tie-up sheet under our noses, and asked "What the bleep is this?"

"Well," replied the dispatcher, "The man got nasty and insisted that I show him tied up someplace when he did not call in. I thought this was as good a place as any." The sheet read "Sadie's Whorehouse, Kadoka, SD." The telephone number shown was a dial-a-prayer service number.

"That's about what I thought," muttered the superintendent as he stomped away. "That cocksucker will need dial-a-prayer by the time I'm through with him." He had evidently called the number out of curiosity.

Brick, But Not An Outhouse

A few years before this the chief dispatcher and I had a little agreement. Every time the agent in the neighboring station, one much larger than mine, had to be absent he would send me there, some newbie to work my station. The pay at this station was quite a bit higher and I had a clerk there. I liked that. In return I was to make sure the work at my own station was done correctly. I was to try to teach the new man what I could. This proved to be a good deal for all three of us.

One year while I was there, a beautiful new brick church was being built. The polished brick they were using came in by rail. They were very selective, to say the least, as to the condition of the brick. I think some of it was flawed right out of the kiln and some was chipped in handling.

As a result I had quite a stack of returned brick stacked behind the freight house. This same petty official came along and insisted I stack this brick inside. He said it was valuable and might get stolen. I told him no, I was not going to do that. I did not have room on the ground level. And the top level, the same height as the floor of a box car, might collapse under all that weight.

A few days later the man was back. Around and around we went. He told me to move it inside, or else. I told him it would be a cold day in hell.

The next week I was back working at my own station when he stopped in. He got more and more abusive, informed me that the regular agent was moving the brick. Around and around we went some more.

I don't remember exactly what he last said, but I do know this was the only time in my forty-two-year carrier that I totally lost my temper. I told the man that I had heard enough. If he did not shut up, I would throw him out of the depot. He did not choose to quit while he was ahead and found himself sitting in a bed of sharp cinders behind the place. He left, screaming that I had ruined his new suit and he was going to see the superintendent.

After I cooled down a little, I thought it might be best if I told the Superintendent what happened. I managed to get the superintendent's clerk on the telephone. He told me that the superintendent was not available at the moment and asked what I wanted. I told him. He laughed, telling me he would inform the "old man" as he called him, and have him call me back.

In a few minutes my phone rang. The first question the superintendent asked me was if I had hit him. I replied I had not, but that I sure as hell felt like it. "Well," he assured me,

"Then it maybe isn't quite so bad. At least you were honest enough to tell me what you did. I'll see what I can do to hush it up."

The following Monday morning the telephone rang. "Have you seen anything of that sonovabitch this morning?" I was asked. It took me a moment to realize it was the Superintendent. I knew exactly who he was talking about.

"No," I replied.

"Well, he is supposed to be headed in that direction, and I want to talk to him. Have him call me if you see him. I just found out the bleeping freight house collapsed over the weekend" The man did not show up that day, or, for that matter, quite a while after that.

The agent at the neighboring town was sort of a worry wart, tight as the bark on a tree. I imagine, with the help of his clerk and the track maintenance people, they had stacked all the brick inside.

Thank goodness for rotten pilling, I thought to myself. This proved to be the last I ever heard of the incident, which suited me fine. I managed to fix the screen door I had shoved the man through. I felt I was even with the world.

Half And Half

It did not take me too long to realize that the chief dispatcher was somewhat like a mother hen with a bunch of chicks. He tried to care for us the best he could, keeping the heat off of us whenever possible. This, however, did not prevent him from chewing on us if he deemed it necessary. But, he made it quite clear if any chewing was going to be done, he was going to do it. We were his boys and could do no wrong in the public eye, if he could help it. This gave us a strong sense of loyalty to him which, I imagine, is exactly what he wanted.

Not too much notable happened those few months that I

spent as a relief dispatcher. I managed to survive my first derailment, which I will always remember. I will also remember a very cold night when a bad wheel on a southbound train picked a switch and derailed one end of the car. Thank goodness it happened while two trains were meeting, both going very slowly. The engineer of the train going through the siding saw it happen, and told the engineer with the derailed car. He stopped at once, and there was no track damage.

The area was joint track, used by three railroads, and I soon had trains stacked up like cord wood. Help was hard to come by; nobody could seem to get anything started. I had a Rock Island road master and a section man at the scene. Then a big side-boom cat, loaded on a semi trailer showed up, but the cat would not start. Somebody from the car department showed up, in his own car, with no equipment.

There was a passenger train sitting a couple of stations away from the derailment. The conductor came to the phone, informing me that the water in the boiler heating system used to keep the coaches warm was getting low. The water supply at this station was frozen. He could not take water there. He said that the thermometer on the wall of the depot read fifty-two degrees below zero. There were over three hundred passengers on the train.

There was a big city about thirty miles from the site. There were plenty of hotel rooms available, no problem. Transporting the passengers was something else, nobody wanted that job. Everybody was afraid that their equipment would not start and that drivers would not be available because they could not get their cars started. The chief of the volunteer fire department in the small town informed me it was his duty to protect the village, not to provide water for trains. The conductor, a nervous tick wired for sound, now a very worried man, reported that the water level in his boiler system was dropping faster than usual. This was due to the extreme cold.

A nice mess. He had every right to be worried.

Then the road master reported that the cat had started, asking me what I wanted to do, nobody was showing up. The derailed car was an empty. I asked him if the cat could tip the car over and shove it out of the way. He said he thought it could, but he had no authority to do that yet. I told him I did not have the authority to let people freeze to death, tip the damn thing over and get it out of the way. I imagine the man was half frozen to death, and wanted to get out of the cold as quick as he could. He did not argue. The cat skinner must have known exactly what he was doing; I do not think that it took him over ten minutes to get the job accomplished.

As a guess, I would say that tipping the car over and moving it was the road masters intent from the start. I can think of no other reason that he had the cat there. The conductor was a very happy man; he could now get water where he always did. The crew on the train sitting in the siding maybe was not too happy, he was held there for two more trains, including the passenger. My car, to my surprise, started after I got off work, so that made me happy. I do not know what railroad the empty car belonged to, but I do not imagine they were very happy. Nobody ever second-guessed my decision. Since this has been forty-five years ago, I do not expect anyone to call me tonight to chew on me.

Not long after I became a train dispatcher, my fears came true. I was displaced from my agent's job. I was spending about half of my time as an agent, the other half as a train dispatcher. The only agent jobs I could now hold were too far away from the dispatcher's office to commute. I chose to go back on the extra board. This resulted in me being home only when I was working as a dispatcher or was lucky enough to work a station close to home for a couple of weeks. The chief tried his best to keep me within a comfortable distance of the office, but too often could not.

It was during this time that I came across the most unusual train order I ever saw. It read “lookout for automobile in a tree,” and then gave the location. The railroad ran underneath a cliff at this area. The highway was on top. Someone had lost control of their car due to ice. It plunged over the cliff and lodged in a huge oak tree. I understood nobody in this car was hurt, but I would not have wanted to be standing on the running board of an engine passing under it if the car decided to fall down.

We dispatched Rock Island trains over a small portion of our division. A Rock Island operator had the chore of providing information on the trains, including the last names of the engineer and conductor. We had a company phone to reach this operator. It often did not work very well. People sounded like they were at the bottom of a barrel.

One day this operator came in on the phone and told me that the afternoon passenger train was called on time. There was a new crew on it, the engineer was crazy and the conductor was wilder. For obvious reasons I questioned this. I asked him to repeat what he had said. He did, and I again understood the same thing. “Well, I don’t want to doubt your word, but what are their names?” It turned out that the engineer was Swazi and the conductor was indeed named Wilder.

One Sunday night I came to work and found we had a new operator at the terminal to the North of us. The dispatcher that worked second shift told me that this operator had only about a week of experience, and he asked me to give this guy whatever help I could. This position was one someone with very little experience should not even think about working. The work load was very heavy. There was a dispatcher to work with at every point on the compass, with many train lists to send and copy. He was the only man available. It was either him or nothing.

I managed to get him in time to issue a set of orders for one of my trains. He was a bit slow and hesitant, exactly what I expected from a very new man. I had no problem with him at all.

Time went by. I got no report from him of the train leaving. Thinking the train should have already departed, I called the crew on a line radio. They told me what time they had left but questioned the absence of a train order covering a slow condition. They did not think there had been anyone working over the weekend to repair it. My records indicated this order was still in effect, and they had received it. I asked the crew if they wanted me to give them a copy of it over the radio. They said no. They had a copy of the slow orders from their previous trip.

It did not take very long at all for my city phone to ring. The new operator had heard the radio conversation and freely admitted what he had done. Just as I had suspected, he was in too much of a rush to fill out the clearance form. He had read the order numbers off of the stack of orders he had. This was something he should not have done. When the crew showed up, he hurriedly filled out the clearance. Two orders stuck together and he missed one. I told him not to worry about it too much. I would talk to the train crew when they got in.

The train crew came into my office, wondering what we should do about the situation. I told them I would as soon forget about the incident if they were willing to. I did not like to go to standard investigations. Nor would I like to see them have to explain why they had not read their orders before they left the terminal as required. At that point, they were very willing to agree with me.

I got busy and forgot all about the incident. About an hour before I was to get off duty the city telephone rang. It was the new operator. He got off work an hour earlier than I did. I told him I had talked to the crew, everything was ok, and to forget about it. He thanked me profusely.

As beat up as he must have felt by that time, I did not have the heart to chew on him. But I did think I would tease him a little. "Nothing to thank me for," I told him. "I am just being selfish. If they fired you, I shudder to think of whom they might replace you with." I was laughing when I said it and he knew that I was teasing him. He had a very hardy laugh. I felt I had made a friend, but little did I realize how often my own words would be thrown back at me in later years.

CHAPTER FOUR MY OWN JOB

A permanent dispatcher's position came open in a town about two hundred miles away. We owned the home where I lived. We hated to give it up, but came to the conclusion that we would be better off moving. I asked the chief if there would be any hard feelings if I took the job. He informed me he would be disappointed in me if I did not take it, and wished me luck.

I now had a new territory to learn and new people to work with. This was a much shorter territory than I had with more traffic. It did not take me too long to make the adjustment. The men I worked with now comprised of a graduate journalist, an airline pilot whose health prevented him from flying commercially, and a man who was well on his way to becoming a doctor when he discovered blood and sick people were not his thing.

This made me feel something like a mental pigmy, a peon among nobility. The latter man worked both the dispatcher's and the chief dispatcher's job. The subdivision was not deemed big enough to warrant a full time chief.

Getting To Know The Chief

I had not worked here very long before I did something that was not too smart. I think that I left some cars sit instead of taking a bad delay on a train. I was behind with my work that morning, and was late getting out of the chair. The chief had not even had a chance to look everything over before the telephone rang. It was the superintendent wondering why I had done what I had.

"Well, hindsight is always twenty-twenty." He then went on to mention about half a dozen rather plausible things that I would never have thought of, favoring the decision I had made. The chief concluded the conversation by asking the superintendent if he didn't have anything better to do than worry about petty shit like this.

The superintendent, who was on a speaker phone, did not sound too convinced and ended the conversation with a grunt. The superintendent was in poor health. He was about to retire and was noted for having the disposition of a junk yard dog. The chief turned to me and said "Kid, don't ever pull that stupid stunt again," laughing as he spoke. I felt I had found a friend.

I had been working there about two weeks when the chief quit reading my transfers. He would check off everything blindly and sign the transfer. Once in a while he would ask me where two trains were meeting and that was all. I made some comment on this and he told me, "Kid, I know how you think, I don't need to read the damn thing."

I also found out shortly this chief was a railroading genius. He thrived on a steady string of nasty black cigars, blacker coffee, and profanity. He kept an empty metal wastebasket near his desk. The basket was big and sat on concrete. Any time he got a phone call that he deemed unnecessary or otherwise undesirable he would drop the receiver into the wastebasket. One morning he got a call. The party at the other end wanted help tracing a car. "Call car scope," he advised them.

"This is car scope" came the reply. Since the only reason for the existence of car scope was to trace cars, the receiver found its way into the wastebasket very quickly. This was after the caller listened to an expletive or two. I asked him why he did this instead of just chewing on the caller for bothering him. He said that this was easier and quicker, and they caught on just as fast. I imagine they did, maybe even before their ear drum returned to normal.

It soon became obvious to me there was a much different tone about this place. Anyplace else on the railroad I had ever been, the trains seemed to get over the road in spite the train crews. Here, they seemed to get over the road because of them.

All of the chief dispatchers I had ever been around had at least a healthy respect for, if not feared, the officials. Here, the officials seemed to fear the chief dispatcher.

Sweeping It Under The Rug

The train crews would do most anything the chief asked of them. Damn all rules and union agreements they seemed to mean very little. One morning I had issued a wait order on one of the regular trains for a local to run against. The engineer on the regular train was the senior engineer in that district. He prided himself on being as close to the time the wait order specified as he could. He never seemed to grasp the concept that the times shown were not meant to delay him. He was not expected to be there at that exact time. He would sometimes swear at the dispatcher, thinking the dispatcher should know better than expect him to be there so soon.

Several times were specified, each at a different station. The rules required the train using this wait order to be clear of the main track five minutes before the time specified at that station. The chief had not been on duty too long before we heard an unusual conversation on the radio. I knew there was something wrong. A moment later we lost our dispatcher's phone. The chief swore a little, soon the engineer of the regular train informed us he was, as he put it, all over the ground.

The chief turned to me and said, "Go on home, kid, it wasn't your fault." When I came back to work the next night, I learned the train not only was on the ground, but there were box cars in the river, on the bluff, and stacked up in a heap. The derailed cars had knocked out our dispatcher's phone. I was glad to see the mess was cleaned up before I got there.

When the chief got there the next morning, the first thing I did was ask him what had happened. He told me the engineer on the local had his watch stop on him. The train crew did not bother looking at theirs. They had tried to get one station too far. The two trains saw each other at a distance. The engineer on the regular train had dynamited his train, knowing full well he might derail it, or maybe he hoped to. The local had managed to stop and back into the siding where he should have stayed in the first place.

Both crews told the chief what had happened. They concocted the story that the engineer on the regular train felt he was running too close to the time specified. Not wanting to be too early at the next station, he touched the brakes lightly. There must have been a "dynamiter" in the train, causing the train to go into an emergency mode of breaking. This was shown to be the official cause. A thirty or so car derailment had been swept under the rug. I fully understood why the train crews did as he asked. He had just kept some of them from getting fired.

The old engineer involved in this incident was a story by himself. Nobody knew how old he was. He was a World War One veteran. He had no birth certificate and no one was ever able to find any record of his birth. His wife was deceased, and little besides his work enthused him. Engineers were required to retire when they were seventy. He somehow managed to get his records changed, not once, but twice.

If one put paper and pencil to it, he must have gone to work for the railroad about the third trimester of his mother's pregnancy to be the age he now claimed to be. He was still an excellent engineer, so nobody seemed to care how old he was. He had a son that was a yardmaster. People joked that the son would retire before the father did. He did not miss it by much.

One morning we were moving two extra sets of engines on his train. He came out of the roundhouse on the sixth unit, which would be in the lead when he left town. He had a head brakeman making his first trip. There was a puzzle switch off the roundhouse lead track that could be lined either for the yard or to the main track. The new brakeman lined the switch wrong and gave the engineer a big come ahead signal. Come ahead he did until he coupled on, rather violently, to the rear end of a passenger train that was sitting by the depot, loading passengers. There were a number of officials from Chicago having breakfast in the dining car. They were suddenly wearing much of the breakfast they had intended to eat.

There was no way to sweep this one under the rug. The subsequent investigation exonerated the engineer. He had done nothing wrong, but he got a lot of good natured teasing about the incident.

We had another engineer also beyond retirement age. His wife, too, had been dead for a number of years, but he did manage to find another interest. It came in a bottle. He was the engineer on the morning meat train, and was famous for the precise stops he made. The meat train was late one morning. There was a bunch of officials in town. They decided this was a good morning to see him perform they were up anyway. He came barreling by the yard office at his usual high rate of speed, stopping as smooth as glass, precisely where the yardmaster had instructed him to.

He than made a rather bad mistake. Seeing all the white shirts, he wondered what was going on. He decided to get off the engine and investigate. He did ok until he let loose of the last grab iron on the engine. He collapsed in a heap, too inebriated to get up. This also could not be swept under any rug. It was the last time he ran an engine. He retired rather than bother attending the following investigation. Unfortunately, now the bottle was his only interest. He lived less than a month.

One night a train had some problem which would delay it a good long time. I had an opposing train that would get a huge delay if I were unable to change the meeting point. I had just let a night operator go home. His shift was up. He lived far away. He was the relief man who worked the job two days a week.

I called the regular operator, no answer. I was busy, and without thinking, I called the old agent. He was the same one I had I had so much trouble reading on the telegraph wire. He answered the phone. "Good morning Ed, I need you to copy a train order," I tell him.

"Be right down Joe," he replied.

Two minutes later my phone rang. "Christ, Joe, I can't come down, I'm retired."

I told him how sorry I was to have disturbed him. It was the last time I ever talked to him. It was a reflex action on my part to call him. He had only been there for fifty some years. I guess I was used to him being there. Awakened from a sound sleep, his reflex action was to get down to the depot. I guess he had done that a time or two before.

I was able to get the regular operator when he got home a few minutes later. If he had a beer or two under his belt, I did not worry about it. He could copy train orders better under the influence than most guys could sober.

We had twenty old refrigerator cars stored on line, awaiting disposition. One day we received notice to pick the cars up on the first Eastbound train that could conveniently handle them. They had been sold for salvage. A few days later we had a train short enough to handle them, and I instructed the train to pick them up. The cars must have been there a while. The crew came to the phone, reporting that the trees that were growing between the cars were so big they raised the cars up in the air when they pulled on them and would derail them if they persisted. We called the boys with the chain saws.

A week or two later I again tried to pick the cars up. There was no problem until the train had gone about half a mile with them. Then every journal box on the twenty cars started smoking, a condition known as a hot box. These cars were built long before the roller bearing was common, and used large brass bushings instead. Each of these bushings weighed several pounds, and brass was rather expensive. Some ambitious thief had jacked each car up and stolen all of the bushings. The journal boxes had become so hot that the grease used to lubricate the bushings was smoking. There was another station only about three miles away. The crew managed to get there without burning off any of the journals, but several were blazing brightly. This time we called the car department.

The Chief

One morning, as the chief and I were making a transfer, George, the assistant Superintendent walked into the office. The telephone in his office started to ring. He commented it never seemed to quit. He was a man getting on in years, not in the best of health, very good hearted, and people took advantage of him. He had no clerk to screen his calls and the chief felt sorry for him. "I'll fix that," the chief told him. By afternoon the division lineman had installed an extension of his phone in our office.

The following morning just after we had completed our transfer, the extension started ringing. The chief answered it. It was the superintendent of an adjoining division. He knew the chief's voice and could tell he was on a speaker phone. It became rather apparent he did not like either of these facts very much. "Where is George and what is he doing?" he asked.

"I don't know, replied the chief, but he is probably out in the yard, making our living for us. What are you doing?"

"What I am doing is none of your fucking business," the superintendent told the chief.

"I know," replied the chief in a very pleasant voice. "I just thought that if you wanted to know what George was doing maybe he wanted to know what you were doing and I could tell him." There were a few muttered cuss words, and the connection went dead.

By that time, we knew where George was. He had been in the yard office and overheard the conversation. He now in our office laughing so hard that tears ran down his face. The deluge of telephone calls dwindled down to a trickle. This mother hen included assistant superintendents among his chicks. The chief seemed to take very good care of them.

He was also very good at appropriating anything we needed. Everybody on the railroad seemed to be short of ties. We had more than plenty. The top row of ties seemed to disappear off a good number of carloads moving through town. The weeds were full of them.

I remember a solid train of close to two hundred grain empties moving through town bound for the harvest out West. By the time he had preformed his magic, I was able to put the train into a ninety-car siding with room to spare. Maybe somebody someplace was short on empties, but our elevators seemed to get them.

We were always short on engines on the branch line that fed our subdivision. The trains overloaded. He appropriated an engine we set out at the end of the branch. We sent it back on the next train to do the same thing again. The engine needed some sort of major service we were not able to perform locally. It had to be sent to a roundhouse on a neighboring division to get it done.

A guy in the control center spots the engine on the roundhouse roster, figures out where it came from, and calls the chief. He told the chief he could not steal power without getting caught. "I guess you are right," says the chief. By then, the two engines he had since appropriated were making their third trip across the branch. Since winter was coming on, he thought it would be nice to have a spare.

Shortly after midnight I always gave a line-up of trains expected to the next division. This was for the use of the yard, dispatcher, roundhouse, and whoever wanted to know when and what was coming. I had shown the morning meat train arriving at 7:45am. The chief was already there when the roundhouse foreman came in on the phone and asked in a very sarcastic voice, "It's seven forty-five kid, where is my train?"

"Look out the window," I told him. There was silence for a moment.

"I'll be damned, he is coming," he said.

The chief got a chuckle out of it. For the first time in my life, I felt like a train dispatcher.

There were two typewriters in the office. One was used to do general work, copy lists, issue messages, or whatever. The other was used to make copies of slow orders. It was almost impossible to take out a thick pad of the onion skin forms and get them back into the typewriter without ruining them. We left them in the typewriter and pecked away at them as time permitted.

The typewriter used for the orders was of ancient vintage. If someone had purchased it at a garage sale for a buck and a half, he would only have done so because he thought the scrap dealer would give him two bucks for it. It was very heavy.

An efficiency expert came along. He did not like the idea of one man using two typewriters. Over the chief's protests, he confiscated the old machine. This proved to be a bad mistake. When he got off work, the chief paid a visit to an officer supply house. He bought a typewriter, not any old typewriter. He was pissed.

The next morning our clerk had the most expensive electric typewriter money could buy. State of the art. All the bells and whistles. Plus all the accessories available. She was a happy camper. She should have been. She had a better typewriter than the secretary to the president of the railroad did. Her old machine now sat in our office.

George approved the bill and sent it in for payment. Someone questioned the purchase. An ordinary typewriter cost about a hundred and fifty dollars, this one cost seven hundred and change. Why did the chief need such an expensive machine, they asked. "Oh," said George. "He did not buy it for himself. Some efficiency expert came along and took a typewriter he needed. He had to take the one the clerk was using and bought her a new one to replace it."

Nobody wanted to incur the chief's wrath over such a minor issue. They chewed on the efficiency expert instead. The chief had his own ideas as to what constituted efficiency.

The Same Old Superintendent

We now had a new division superintendent. They promoted the man I had previously worked for to this division, which was a more prestigious position. I had the same boss again, a fact that made me glad. The man was gruff, but had always treated me fairly. The known devil is always better than the unknown. I came to find out this man and the chief had once worked together as train dispatchers, they were personal friends. At this time the superintendent had missed giving a train an order. The chief discovered it as he came to work. They managed to get the order to the train and prevented what would have almost surely been a head-on collision.

This superintendent made it a habit to call the chief a little after eight each morning. He would, at times, complain about the train operation. Everybody knew what the problem was. Over the years the trains got longer and the sidings did not. We now had too few sidings long enough to hold the trains we ran. We sometimes had some bad delays. The chief would let him complain for about so long, and then remind him of the fact that when he worked there he sometimes had the trains meeting in the cornfield. At that point it did not take the superintendent too long to find something better to do than talk to the chief.

I soon took to hanging around work until after this call had been completed. I was not only getting a free lesson in railroading, but I enjoyed the two men bantering back and forth.

I could not understand it. This chief seemed to violate any rule he desired any time he wanted to. After we ran detours on our railroad, he thought nothing of tossing all the train order books into the furnace. If there was no fire in there, he promptly made one. All of the evidence of our short cuts, sins, both mortal and venial, became ashes. He did not seem to hesitate to chew on any official when they did something that was not to his liking.

The rules examiners never seemed to notice the missing books. They wrote up a laundry list regarding how he kept his train order book. He got a nasty gram from the powers that be about it, and his reply was always the same and very short. Exceptions acknowledged. He once got a particularly nasty letter from the chief rules examiner demanding an immediate explanation of why he had not signed the train dispatcher's transfer of orders on a certain date. He scrawled "because I did not read it" across the face of the letter with a magic marker and mailed it back to him.

I asked the day yardmaster, who had been around a long time, if he had any idea of how the chief got by with all this without getting fired. He smiled at me and said "It would only take them a day to fire him, but they could not replace him in a thousand years, and they all know it. And he knows they know it."

How true these words were became evident to me some time later. There was some sort of an issue with the switchmen's union that worked the yard at division headquarters. The switchmen staged a slowdown strike. I do not remember what the issue was and would maybe not want to go there if I did remember. These switchmen were known to be among the best on the railroad. Now they crossed all the T's and dotted all the I's. Instead of moving through the yard like a mad bunch of bees, they moved like a pregnant snail through a thick pool of molasses on a cold day.

These engines ran right by the window of the superintendent's office, and watching them was driving him half insane. He threatened to make a ghost town out of the yard and do all the switching elsewhere. He called the chief just as we were making a transfer one morning, raving and ranting about the situation. "And I think I can do it, too, if I can figure out a way to keep two of your trains out of this damn yard. I've been working on it for days. I'll be damned if I can figure out how to do it without crucifying the meat train, which I won't do."

"That's easy," replied the chief.

"Easy! What the fuck to you mean, easy?" growled the superintendent.

"Sure, all you have to do is get the stupid ass hole working your old job to reverse the blocking on the two connections. If he knows how. The meat train never gets here until after six. We put the seven o'clock engine here to work an hour early. The south cars stay here. We take the east cars off the south train and give him the south cars. The east cars off the south train can move on the afternoon meat train. Then we switch up the east shit the afternoon meat train leaves here. It can go on the meat train the next morning. Then nobody has to go into the yard there."

The superintendent was not convinced, but very interested. "But that means the meat train will have 50 or 60 cars to set out and about that many to pick up. That will take a lot of time."

"Not really. I'll have George arrange for the yard to have the fill all lined up on number one track with the air all made and pumped up. When the meat train gets here he should have 30 or so East cars on both ends of his train. We use the midnight job to grab onto one end of the train, the morning job onto the other. We shove em onto the fill on number one, make two joints, and the train can go. Shouldn't be here over about twenty or twenty-five minutes, a half hour, tops."

"Just might work, at that." "Start Friday?" the superintendent asked.

"Your party, dance whenever you want to" the chief assured him.

That morning George was on hand bright and early to make sure everything was lined up right. He left with a smile on his face. The train was in town twenty minutes. The normal time it took him to pick up a few cars there.

The bean counters said keeping these trains out of the yard saved the railroad over a million dollars a month in switching costs and delay. The superintendent got credit for a brilliant move, and if I remember right, won an award for being the official of the year.

The chief was happy to see his buddy get the recognition, and commented it would be nice if they got rid of the bean counters and used the money saved to build some longer sidings. I do not know what happened to the switchmen's dispute. I do know the yard went from being a major terminal to just a wide spot in the road.

As far as the officials were concerned, the chief had two bad habits. He always seemed to be right. The other was he seemed to have the solution to most any problem they were about to have before they even knew they had one. If they defied him, he would watch them paint themselves into a corner, then gleefully serve them more than adequate portions of crow.

The chief's work load would have driven an average man to his knees over any extended period of time. He thrived on it, seemingly enjoying it, as though it were some sort of a game of wits. And to him it maybe was, one that he normally won.

The rules department got all over the chief about the condition of the train order books. They had a light cardboard backing. He preferred sitting on four of them instead of the cushion the rest of us used. When they got rumpled from the abuse he went and got four more train order books. He commented that the rules examiners who checked them probably would not get their noses nearly as close if they knew how often he farted.

We copied our own train orders. We placed a carbon paper under the train order form to put a copy of the body of the order into the book. Five or six orders per page were about normal. The chief thought nothing about using a whole page for an order, the whole body of the order running at a slant. He maybe should have become a doctor, his handwriting was next to impossible to read.

I once asked a trainman how he could feel comfortable leaving town with orders that looked like that. The trainman informed me that all he needed was a match book with a little advertising on it. It got so I could figure out the chief's writing. I guess the conductor could, too, but it took me quite a while.

Some head bean counter wanted some report of some kind and called the chief asking for it. The chief promptly told him he was too busy. He tried to call George, but would get the chief on his extension. The bean counter wrote a series of letters, each nastier than the last. The last was a real scorcher. That got him a reply, not that he liked it, but he got it. It read "I work the chief dispatcher's job, the dispatcher's job, the operator's job, and sometimes the yardmaster's job, all at once. I try to do the best I can, and if you find anybody dumb enough, including yourself, willing to try to do this better, I am more than willing to step aside and let them try. However, until that time, shut up."

The chief had turned down promotion after promotion. I think he could have gone anyplace on the railroad that he wanted to. He preferred working his eight hours, going home, reading his Wall Street Journal, maybe having one off the top shelf, and relaxing. To him, relaxing often meant tending to his huge lawn, garden, and fruit trees. I am sure that he was rewarded far more than any official who spent a great deal of his time at derailments and crawling through the weeds to make signal tests. His understanding of the stock market was about on par with his ability around a railroad.

We had a third shift yardmaster that could not get the place warm enough in the winter and cool enough in the summer. The thermostat was in the yardmasters office. I kept the office window open all summer to keep from freezing to death, and open all winter to keep from baking.

George once told the guy to let him know if he ever could get any more heat out of the furnace. It seems that he had a turkey he wanted baked. He would have his wife stuff it, bring it to his office, and it should be done by morning. It was easier opening the window than arguing with the guy.

One night we had a party, well, sort of one, anyway. A yard clerk who worked third shift was retiring, this was not only his last day of work, but also marked the day he would have sixty years of service. We chipped in, bought some goodies, and there were several people on hand when he came to work about eleven thirty. He was dressed in a suit and tie, wearing a nice hat, neat as a pin, his normal attire.

He went through the messages on the desk, then his own personal message box. Finding nothing, he turned to us and said, in a very sad voice, "It's not here again."

"What's not here?" someone asked him.

"Well," he said, now almost looking like he was going to start crying, "When I started here sixty years ago, they told me it was temporary. Nobody ever saw fit to notify me I had a permanent job, so now I will have to go through life without one." Then he started laughing, and needless to say, so did we.

One night I coded in a line radio with the intention of checking to see where a train was. I heard the conductor holler at his engineer. "Pu pu put them in the big hole, Karlee. Wa wa we are on the ground". The man was so excited he could hardly talk. I can not say that I blame him.

The car just ahead of the caboose had derailed. One of the derailed wheels was running at an angle to the train. The conductor had the bay window open, checking his train for signs of trouble as they rounded a curve. The derailed wheel was throwing a steady stream of gravel which was striking the conductor in the chest. It was dark and the conductor could not tell how bad the derailment was. As it turned out, it was only one end of one car, and they were able to stop before any further damage was done.

One night the meat train derailed one end of a car. It was winter and below zero. The car traveled fourteen miles before it re railed itself on a crossing. The track patrolman reported this when he inspected the track. The car department inspected the whole train and was unable to tell which car had been derailed. I would say that this train crew was not checking the track behind them very often. There was snow on the ground and the marks the car made were very visible. It was this sort of thing that helped end the era of the caboose.

We were short on engineers. One of our engineers had a brother in law that had made a few trips with him, with the idea of getting a job. I do not remember if it was George or the traveling engineer that hired him, maybe I never knew. Hooray for whoever did. Something happened and they were forced to put this guy out there without much of any further training. He had just passed the required exam. On the return leg of his first trip he was presented with a two hundred and thirty five car train, a real monster. There were many long cars in the train; I think it lacked being three miles long by about two inches.

He managed to not only handle the train very well, but made track speed with it. I could not believe it, and still marvel at it forty five years later. His conductor told me that he handled the huge train as well as the average engineer with twenty years experience could be expected to handle a train a hundred cars smaller. He just had a natural ability that very few possess. He was also one of the most pleasant people I have been around, liked by all.

George liked him very well. He had no reason not to. He came into my office one morning, asking me how the new engineer was doing. "Fine", I told him.

"I think he is going to make a good one", George commented.

"Wrong", I told George. George looked very surprised at my comment. "He already is a good one". I continued. George beamed.

However much ability this man had, I was still a little concerned about him, especially when it was foggy. He was, after all, a new man. I would ask him where he was quite often, more than anything to see how he was doing and if he knew exactly where he was. He always did. One morning he answers me "Going by the tie pile". I thought he just forgot to tell me in which town. The next time I asked him, I got the same answer, which I thought odd. On his next trip I asked him again. Same answer, but I could tell he was laughing. He sensed I was trying to baby sit him a little, and this was his way of telling me that he did like it too much or need it. He was right.

I saw him some time after he had retired. I asked him if he remembered that morning. "I sure do, Joe, I was scared shitless" was his reply.

I have news for him. So was I, and I was not even riding the train.

The terminal at the West end of our territory belonged to the neighboring superintendent. We operated a local out of there. The wages to pay this crew came out of his budget. There was a long, steep hill part way over this territory. The local would take about a thousand tons of through business up this hill every day, in addition to doing the local work. The superintendent decided to cancel this local to make his budget look better. He got his boss in on it. There was a conference call about the subject. The chief told them it would never work. They cancelled the local anyway.

Some days later there was over ten thousand tons of East business sitting in the terminal occupying valuable space. There was local business sitting in many places on the line. The through freights were not able to even come close to keeping up with it. The telephone rings. It was the superintendent of the neighboring division and his boss. "What are you going to do with the ten thousand tons of East business you have sitting?" asked the Superintendent.

The chief replied, "Those are not my ten thousand tons, they are your ten thousand tons, you stupid son of a bitch. I told you taking off the way freight wouldn't work. Now go fix it, get off the phone and quit bothering me."

The next morning I made it a point to hang around for the morning call from our Superintendent. I had heard about the incident, and saw a notice up about the local being reinstated on Monday morning. I was curious to see if there would be any repercussions.

"I hear you called my replacement a stupid son of a bitch yesterday. Is that right?" was the first thing that he asked.

"Well, yes, something like that, anyway," replied the chief.

"He called me yesterday. He seemed pretty pissed off. I guess the boss ate his ass out over the deal. I promised him I would talk to you about it, so I have. In my opinion, you were flattering him, but I didn't tell him that, didn't want to hear him cry."

The local lived far longer than the man that tried to kill it. After his death, one of the guys who had worked for him remarked that the only reason he had not stopped by to piss on his grave was that he hated to stand in line.

A Bastard Learns A Bitter Lesson

Having some time off, I happened to be in the area where I had started dispatching. I stopped by my old office to say hello. It was March, just after a bad snow storm. The temperature at that time was near the freezing point, then it cleared up and became bitterly cold, several degrees below zero. There was a branch line South of the office running almost straight North and South. A very strong West wind accompanied the storm and managed to fill the many deep cuts on this line full of snow, in spite of how heavy and wet it was.

There was a snow plow working on the line. The progress made was very slow. There was twelve feet of snow in many places. The extreme cold had turned it rock hard.

The superintendent, evidently parked at some vantage point, was very unhappy at the lack of progress the snow plow was making. He started berating the engineer. The engineer would go ahead until he stalled, then back up a little and repeat the process. The plow had a heavy load of soybeans behind it, placed there as a buffer, to protect the two engines in case the plow would derail.

The engineer, maybe not the fastest rabbit to ever come down the bunny path, was very reliable. He was as steady as could be, a very experienced man who had a pronounced Scandinavian accent. He had preformed this chore a hundred times, or maybe hundreds.

Each time the superintendent criticized him, his answer was the same, a very respectful "I am hitting them as hard as I dare, Sir."

Each criticism was a little harsher. The final one ended with "Back up and hit those snow drifts like a man, if you know what one is."

This time the engineer's reply was a "Yes, sir," that

dripped venom.

The dispatcher on duty and I looked at each other, both of us saying "Oh Oh," at the same time. The chief, who was in the next office with his door open, made several derogatory remarks as to the Superintendents' linage and intelligence. He reached for the telephone to call the car department. They would be busy very shortly.

All was silent for a bit, then the engineer's voice came in on the radio. "Vas that hard enough, Sir?" The snow plow now sat near the right of way fence to the left of the track, the carload of beans was tipped over to the right, and the first unit lay on its side to the left of the track. The lead trucks of the second engine were derailed.

The superintendent, now afraid that the engineer may have been injured, asked him if he was alright. "Yes, sir," came the reply, "I'm back here on the second unit, but dare for a vile I vas vishing that I had a turd." They say that a picture is worth a thousand words. The engineer had drawn a very nasty one for his boss.

A Good Boss Goes Bananas

One morning our superintendent walked in about five a.m. I greeted him with a pleasant "Good morning."

"What the fuck is good about it," he replied, walking into the vacant clerk's office and slamming the door behind him. This man had never been anything near warm and fuzzy, but this was not his typical behavior. He had not even bothered taking a peek at the train sheet to see where the trains were. He immediately started making phone calls, sometimes screaming at whoever he was talking to. He was still in there when the chief arrived. The clerk patiently waited to get into her office. I left.

The next week the superintendent started calling me at four o'clock in the morning. He was raving and ranting over things that were very normal and that he otherwise would not have paid much, if any, attention to.

After three or four mornings of this, I had enough. I told the chief about it, saying I knew they were friends, but I had let this go as far as I intended to. I said I would not take it up with the union without his approval, but please talk to the man. There has to be something wrong.

When the chief came to work the next morning, he did not look very happy. "I never had a chance to make that phone call for you," he informed me. "They took him to the booby hatch yesterday morning." He looked like he was going to cry. He was not taking this news well. "They tell me he had a complete nervous breakdown, and that is all I have been able to find out."

Not only was I surprised to see how hard the chief took the news, I did not think they were that close. The breakdown was about the last thing I had expected. This man seemed to be about as hard nosed as they come, seemingly had a good marriage, and no personal problems anybody knew about. He seemed to have it made. He worked long hours, to be sure, but he was used to that. I never did find out what the problem was. Maybe we all have our own personal demons to fight that others are unaware of. It took some time, but he would fully recover.

Whoever took over the superintendent's job was a smart guy. He more or less let George and the chief run things. We heard little from him. In my opinion, the new superintendent could not have done better than having those two. Given a freer rein, George now had more of a chance to prove how fine an official he was. He was not only very capable but a real joy to be around. One of the finest people that I have ever met. I left with nothing but fond memories of him.

We had trouble keeping switchmen. Poor George would hire two and three would quit. At night the yard was not a pleasant place to be. Snow, ice, rain and the mosquitoes were all a factor. Something that a rat found good to eat was always shook loose when the cars were switched. Any large spot of food attracted many of them. When a switch engine came by and disturbed them they moved like a gray carpet. They did not move very fast, most of them were too fat to do anything but waddle.

George hired a pig hippie. His braided hair hung down almost to his belt. He had a long scraggly beard that contained the residue of meals he had consumed days ago. He had tattoos anywhere you wanted to look, and maybe in places that one did not want to look. His clothing looked even worse than he did.

As I came to work one night, the third shift yardmaster was trying to get out of the car someone used to take him to work. He would take two steps ahead, one back, and one sideways. Using the walls of the hallway to steady himself he managed to get to the other side of the building and crawl into the switchmen's shanty. He was about as drunk as a man could get and still remain somewhat upright.

The pig hippy had been there a week, his first man two days, the third man was working his first shift. I took some nasty short cuts when marking the train lists, and lined up the yard work for them to the best of my limited ability. We managed to get the trains through town without killing anybody or derailing anything. The day yardmaster comes to work, notes his less than pristine yard, and then heads for my office. He knew exactly who the culprit was.

"How did you enjoy your first day as a yardmaster?" He asks me, grinning.

My answer was a smile and a shrug of my shoulders.

"How bad was he," was his next question.

"More than plenty. I think he is still sleeping in the shanty." I replied.

"He does this about once a year, and he was overdue. Oh, well, the last time it happened the yard looked worse. If he isn't awake in an hour or so, I will wake him and have George take him home." The yardmaster left, shaking his head. Neither he nor George wanted to see the man fired. He normally left the yard in very good condition, despite all the new switchmen, sometimes having to do the work himself. The next man maybe would not do this.

The first shift yardmaster was a good man with a yard too small for what transpired there. I liked the guy a lot, even if he did have the annoying habit of being a bit grumpy in the morning. When I had heard enough out of him I would tell him seventy-six had half of the box cars in the world, sixty-seven the other half and they were going to meet in his yard. He would promptly stomp off to his office. By the time I went home he had recovered from his blue funk. He would laugh and shake his fist at me. It is unfortunate that my prediction of where the two trains would meet was right far too often.

The chief was waiting to sit down and the pig hippy walked by the window. "What the fuck was THAT, a switchman?" He asked.

"No, no. THAT was no ordinary switchman. He is the foreman," was my reply.

Leaving out the profanity, the chief did not say anything.

However, the laugh was on us. The hippie soon became a genius at switching cars. I could always tell when he was working. The trains left town quicker. He maybe did not know why cars had to be lined up as specified, but just tell him and get out of the way. And he stayed for a long time.

There was a very long car on the head end of the meat train one morning. The train was late, the engineer in a hurry. The long car was sitting on a sharp curve. The engineer tried to get the train moving in too much of a hurry and lifted one end of the car up, derailing it. People were just getting to work. The conductor, old, smart, and experienced, got a lift to the head end of the train. The car foreman arrived. This car foreman was not noted for his ambition. He took one look at the car sitting on the curve and walked the three blocks to my office to call his boss and tell him that he needed the derrick to rerail the car.

The section foreman and first shift yardmaster arrived on the scene. The section tool house was very near the derailed car and it had an old set of frogs in it. He was talking to his boss, giving him all his reasons he could not rerail the car. His boss evidently needed some convincing, they talked a while. I touched the car foreman on the shoulder, pointing to the cars going by the window. He gave me a dirty look and kept talking. This time I was not too subtle. I got right up to his ear and told him, "Tell your boss the car is rerailed and the train is leaving town, that's him you see going by the window". I don't know what his boss said, I turned and left.

Then, the chief lost a battle. Well, he lost it for a while, anyway. The railroad abandoned a branch line and was going to tear it up for salvage. We had used this branch line from time to time to get two huge trains around each other. We would head the Westbound in on the branch line, than have him back out after the Eastbound was gone. The chief wanted them to leave a couple miles of it intact for this reason. No amount of objection, pleading, begging or profanity was enough to keep it from happening. The branch was gone. Not even a switch and a stub of track was left to leave a bad order car there.

Grain prices were poor. People were holding grain. The crops were not as good as usual. We staggered by. Grain prices got better. There was a bumper crop and grain was sold to make room for it.

In the meantime, we got larger engines and the trains got longer and longer. The shit hit the fan. The trains often could not make it over the subdivision without the use of a relief crew. It cost about a million dollars to build a new long siding six or eight miles from where they tore up the branch line.

I was no longer working in that office to see it. I would imagine that there were several more than ample servings of crow passed around. Several years later, they had to build another siding East of the area. They paid dearly for replacing what they had probably sold for a dime on the dollar.

CHAPTER FIVE THE WEST COAST

Heading West

Some bean counter someplace came to the conclusion they could save money by eliminating a set of dispatchers in the office North of us. I got bumped. The only place I could hold a job was on the West Coast. It did not matter that eliminating these dispatchers was not a feasible thing. Trains sat and accidents happened. They had trouble filling the overloaded positions.

It took them several months to figure out this was a very expensive economy indeed. They reestablished the positions, only to find out some of the men had gone to work for other railroads, gone back as operators, or retired. They had more expense and grief training new people. I guess this is another issue, I will get off of my soap box, I have been there too long already.

It was a very long ride. When I got out there, I liked my new chief immediately. Again, I had a new territory to learn. My fellow dispatchers were very helpful. I discovered three of us, now almost two thousand miles from home, had grown up within twenty-five miles of each other. I was allotted some break-in time and had no trouble learning the territory.

I was soon to find out my new chief handled the officials in a far different manner, using humor and good nature as a tool. The Superintendent was not an ex train dispatcher, and unless there was a real problem, pretty much left us to our own devices. I had worked for him before, as an operator, and liked him. It was quite obvious he and the chief got along extremely well. They were both fun-loving, liked to feed each other the raspberry's, laughing, like a couple of school boys.

One advantage this chief had was everybody seemed to like him, despite the fact he was somewhat of a rascal. He was a small man with very long narrow feet. He wore western boots with pointed toes. On the rare occasions he had time to put his feet on the desk, he looked for all the world like a skier who had neglected to take off the front half of his skies.

He had the uncanny ability to predict where his engines would be two days from then. You could wake him up from a sound sleep at three o'clock in the morning. He could tell you exactly where any of the more important trains we ran should be, how many and what type of engines they had, and most likely, the number of the lead unit.

I was working second trick. He promptly nick-named me big hoss and the dispatcher working the other second trick position little hoss. He then bragged about how much his two horses could accomplish, saying that if one hoss could not do it, the other one sure as hell could.

We got a new trainmaster at an outlaying terminal. A local ran out of there each day, on the main line for some distance, then up a branch line and back. The dispatcher working the job would call the crew for duty whenever he felt he could move the train with the least delay. The trainmaster decided this crew was to report for duty at nine thirty in the morning. From time to time this local would not be able to get back within the twelve allotted hours. A relief crew had to be used.

The chief would tease the trainmaster, he would ask him to send the money he was wasting to us, as the dispatchers would love to have a party at his expense. Then one morning the trainmaster discovered he had an empty roundhouse. A train had engine trouble, and we had used his engine to get the train over the next hill. He called the chief, all excited, asking what he should do.

"Well" the chief told him, "I guess you could stick the lead drawbar up your butt and take off. The cars are all empty, so that might not be so bad. Getting the loads back might be worse. I think you will find it easier to let the dispatcher call the crew from now on, though" The chief was laughing so hard he could hardly talk.

The trainmaster, expecting a chewing, and a worse one if the chief took the matter to the Superintendent, was glad to hear the laughter, I am sure. He got to the point where he would ask the chief before he did most anything. Somebody quipped, maybe he asked his permission when he wanted to visit the men's room.

One day a semi hauling one huge log, about sixty feet long and eight feet in diameter, lost his brakes coming down the hill above our yard. The driver somehow managed to get through about forty blocks of city traffic without hitting anyone. He did very little damage, mostly limited to parking meters and rear view mirrors of parked cars. There were quite a number of cars waiting at the crossing for one of our trains to pull into the yard. The semi driver veered to the left and managed to get around them.

The caboose and the semi got to the crossing at the same time. The log crushed the cab, of course killing the driver, and struck the middle of the caboose. The conductor and rear brakeman were in the caboose, and the log passed through the caboose between them and came to rest sticking about fifteen feet out of the far side. The concussion broke the air on the train, which was going very slowly and soon stopped. The impact moved the whole crossing over about two feet, and the caboose was now shaped something like a banana. The brakeman and conductor were unhurt. When the conductor handed in his report of the incident, I felt that he had made quite an understatement. The report concluded with him writing "caboose skun up considerable."

Some New Horses

We got some new engines. We needed a stable of workhorses. What we got were only a few, and they were more like racehorses. They had more than enough horsepower, but too few driving wheels, and they were not a heavy engine. We had a very steep hill coming out of the main yard, and a small yard at the top of the hill. The switch crew would take cars up the hill for the through train to pick up. Our old engines could take 600 tons up the hill. The rail was worn, the top of it being maybe about only half as wide as it should have been. The new engines, with over double the power of the old ones, would spin their wheels as if though they were on ice, with anything over 600 tons.

We were always short of engines, the chief complaining bitterly. He was told that with all the horsepower he had now, he should quit complaining. He told them that he was no better off than he had been, the new engines could not pull any more than the old ones, or maybe even less, and many of the old engines were sent east. They refused to believe him, and flew out to prove him wrong.

Now, the chief did not mind playing games, especially if he could stack the deck in his favor. He arranged for the yard to make up a bunch of cars, loads on the rear, empties on the head end, all having been weighed with the scale tickets to prove it. He had the roundhouse get an old engine ready, tanks full of fuel, every grain of sand it would hold, vital fluids topped off, making the engine as heavy as he could get it. He used a new engine that had just got in off the road, fuel tanks almost empty, about as light as it would ever get.

He asked how many tons they wanted to try. I think they started off with about twelve hundred. Needless to say, the engine started spinning long before it hit the steepest part of the hill. They started setting out empties. When they got the cut of cars down to 600 tons, the engine was able to get up the hill, gaining speed as it went. When it became obvious that they could make the hill, they were told to back down. "Now, he told them, watch this." He then switched engines and had the switch crew back onto two of the empties that they had set out.

They crept up the hill at about four miles an hour, but managed to make it. He had proved his point, but never did tell them that the cut of cars pulled easier with the empties on the rear, or that the new engine, had it been full of fuel, might have been able to do the same thing.

We had an old dispatcher in the office that had some sort of a mental collapse, the man had an insurance and real estate agency that he and his wife ran, and I think that he had spread himself too thin for too long. He was on medication. If I remember right, the doctors cleared him to work the chief's job, which he normally worked on week ends, but not any job where he had to issue train orders. One Saturday, being short on engines as usual, we were trying to figure out how to make eight engines do the work of about a dozen. The answer we came up with was rather complicated, involving a couple of engine swaps and ping-ponging a couple engines back and forth. The Superintendent was listening to us, frowning. "Sounds rather scary to me," he commented, "what are you going to do if it doesn't work?" he asked.

I thought the man's reply was rather novel. "Well," he said, "I don't know what Joe is going to do, I don't know what you are going to do, but I'll take two of my no-give-a-shit pills and put my feet on the desk." I don't know how happy the superintendent was, but he laughed and went back into his office. The plan worked well enough, and the pills stayed in the bottle.

About this time, I heard two bits of news. The operator who had missed giving my train an order had, some time ago, been promoted to a train dispatcher. The other was that the official that I had the row over the bricks with had been transferred to our territory. The chief was rather surprised that I knew him. I told him about my experience with him, and said that we would give him "a West coast welcome."

It seemed that no matter where the man went, he was

urgently needed someplace else, usually somewhere at the other end of the territory. We kept him on the road and busy so much that he must have felt like home and a good night's sleep were but a fond memory. He knew that he was being raw hided. He stopped in to see the chief one day, asking what he had done to us to deserve this kind of treatment. The chief replied, "Nothing, your reputation preceded you, and we were giving you a sample of what your life out here would be like, if you do. If you promise to treat us all decently, I'll call the dogs off." We had no problem with the man, and in fact, he was good at his job, and nice to work with from that point on.

A Holy Chewing

I had an operator who had become a religious fanatic working the job at the entrance to our yard. It was a junction point, trains going in two directions, as well as an interchange point with two other railroads. It was a busy job, sometimes hectic for brief periods of time. This man would sometimes come in on my dispatcher's phone, or call me on the city phone, asking questions like "do you know the bible says," followed by whatever he had to tell me, or want to discuss with me. I did not have time, as a rule, to do more than tell him I was too busy at the moment to discuss it. It got old after a while; he would interrupt my train of thought, and drown out whoever I might be talking to on another phone. He had been a dispatcher for quite a number of years before his preoccupation with religion had resulted in a near-miss head on collision. They took him out of service for a while, then something similar happened, and they forced him back to an operator's job.

One night a foreign line conductor called me on the city phone from the first station the other side of where this man worked. His train was operating as an extra, and should have had orders to meet two opposing extra trains at this station, taking the siding for both of them. He arrived there finding an extra train on the main track with the switch thrown for him to go in on the siding. The engine number of this train no way matched the order he had, it was even owned by a different railroad. For a moment, I could taste brass, thinking that I had somehow fouled things up. I hurriedly checked my records, finding, to my huge relief, that he had been cleared with both orders. I asked the conductor what time his clearance form showed. It did not match mine, and I knew what had happened. When he cleared the train, he read the order numbers from the pile he had, then when the train showed up, filled out the clearance form and missed one order.

The conductor was a friend of his, or at least knew him, and told me that he did not want to see the man fired. I told the man that I would talk to the operator, and call him back. The operator answered the phone and I asked him if he had an extra copy of a train order there. There was silence for a moment, and then he gasped. "Did anything happen?" he asked.

"Not yet" I told him.

"Don't tell the chief." "Don't tell the chief," was all he could manage to say. I told him that I would call him right back.

The conductor answered the phone before it hardly had a chance to ring. "What did you find out?" he asked.

"Well, he missed one order, good thing it was the first one, and not the second," I replied.

"What do you want to do?" he asked.

"Let me give him a talking to, then you take a bite out of him, and we can decide then," I told him. He agreed. About then, another operator informed me that the second extra was stopping. He had brakes sticking.

I wondered how I should approach the man. I knew that if I cussed him out good, which was my first inclination, he would take it as just some heathen using four letter words, and it would do little or no good. I decided that maybe I should approach him from a plane that he understood. I called him, he answered at once.

"Does not the bible say, render onto me what is mine, and render onto Caesar what is his?" I asked.

"Yes, yes," he replied.

"Then I think that you are seriously shorting him." "Does not he pay your wages," I asked him?

"Yes, yes," he agreed

"What does the bible say about stealing?" I asked.

"Oh, no, you are not supposed to do that," he managed to reply.

"Getting paid to do your work and then doing something else is the same thing as stealing, isn't it?" I asked him.

"Well, yes, I guess so," he said.

"And what does the bible say about killing people? You almost managed to do that tonight." "I know you did not do it on purpose, but they would have been just as dead." I do not remember all of it, but I ended up by telling him, "You had better start living your religion instead of just preaching it. The way you do things is no better than a self-professed holy man on his way to a house of ill repute, stopping to preach religion along the way. Now hang up the phone. The conductor wants to talk to you; you almost harmed him most grievously." I called the conductor back, telling him that the phone should now be free; I was through talking to him, and to call me back.

I was soon informed that the second extra train was now moving again. The conductor called me back inside of two or three minutes. "What on earth did you tell that man?" he wanted to know. "All I can get out of him is some blubbering, telling me he is sorry, and that he is a bad Christian." It seemed that I had reached him.

"Plow it under?" I asked the conductor.

"Plow it under, and thanks." He did not need to say more.

We had been very lucky. If he had missed the second order instead of the first, the two trains would have met someplace out there in the darkness at a combined speed of about a hundred and twenty miles an hour. It would not have been good. If I had not been so busy chastising the operator, I maybe could have advanced the train a station when the second train was delayed, but I did not think that the conductor would complain, sitting there for an extra fifteen or twenty minutes. That was small potatoes.

I let the operator stew for a few minutes and then called him. I told him that the conductor and I talked it over, and we thought that giving him another chance was the Christian thing to do, but we both expected him to perform his duties in a proper manner from now on. He threw himself into his duties with the zeal that he had his religion. He must have been one darn good dispatcher. He had been around there for about thirty years; I had been there for maybe three months. Far more qualified than I, he saw things that I would have missed, and seemed to watch over me, like a guardian angel on my shoulder. And he did the same thing with the other dispatchers.

This was the only time in my dispatching career that I took what I considered a serious bite out of an operator. I was confronted with a serious moral issue, and placed on the horns of a dilemma. If I reported him and he got fired the man that replaced him may have made a similar mistake. If I did not, he could have done the same or worse again, but I hated to throw his experience and knowledge away. I just hoped that I had made the right choice. He retired working the job, so I guess I got lucky, and guessed right.

We Play In The Snow

Fall came, and with it, snow in the mountains. It would snow a foot an hour, and snow for a long time. I learned first hand that altitude has much to do with snow accumulation.

There was a siding at the base of the mountains something a little over a mile long. I have seen five feet of snow on the East switch, and not enough snow on the west switch, a scant four hundred feet lower in elevation, to make a decent snow ball. We had to run a rotary snow plow ahead of almost every train. This required the use of two engines, one to push the plow, the other to turn the huge wheel of the rotary plow.

One week end we were forced to use what they call a "mother" engine on the plow. We were using locatrol, which means that there are two engines half way back in the train, governed by radio signals from the lead, or "Mother" engine. Putting all the engines on the head end of the train would pull the train in two. Many of the engines were equipped to receive signals, but not too many could send them. Not only did we appropriate the engine, it got wet, and we had to bring it back to town to dry it out in the roundhouse. This resulted in us having to leave about a thousand tons of East business sit, not having the proper power to move it.

The following Monday the chief sat with his feet on the desk and the superintendent stuck his head in the door. With a big smile on his face, he shook his fist at us and said "You cock suckers." He evidently had been called upon to do some explaining about our using the engine to plow snow. He knew we had to do it, or shut the railroad down, but he had to chide us a little anyway.

The chief, in a very plaintive voice, replied, "Gosh, boss, we're all trying to quit." You could hear the superintendent laughing all the way down the hall.

Shortly after I had started work, I was having a problem trying to figure out where two trains were to meet. I knew what time the crew on one of them was going to go to work, and knew only when the crew of the other was rested and available for duty. The track was slow on the branch line they ran on, it was a desolate area, no operators along the line, and no line radios. There would no way to change the meeting point, once it had been established. I asked the chief if he could help me. We discussed our options, and both came to the same conclusion. We thought the eastbound train would get some delay, but it was the best we could do. The next day, when we got the delay report from the Eastbound train, he showed a ten minute delay. We were still patting each other on the back when we got the delay report from the other train. We quit when we saw that the other train had sat there for about two hours.

One day I came to work and was reading what amounted to our in-house newspaper, containing messages of instructions, cars that were to be moved promptly, and whatever else we needed to know before we started work. There was a message in there stating that effective at once the hottest Westbound we ran was not to leave the main track, anyplace, for any reason. My first thought was that the chief was pulling my leg. Then I thought that if he indeed wanted to pull one of his practical jokes, he could do far better than that. As I walked into his office he looked up at me and said, "Nope, not a joke, just a bad rule." "And the way you get rid of a bad rule is to enforce it to the letter."

This shocked me. There was an area on top of the mountain level enough for a siding. There was a tunnel, about two miles long, just West of the West switch leading into the siding. If an eastbound was allowed to use the main track, he could stop his train at the East switch, on fairly level terrain. If this Eastbound train was to head into the siding, he would have to stop at the West switch, his entire train hanging down the hill.

When he stopped to line the switch for his route, the locotrol engines that he had in the middle of his train would be in the tunnel, where there was no radio reception. These engines had to receive a radio signal every few minutes or they would shut down, and only idle. By the time an Eastbound slowed down, stopped, and threw the switch for his route, that time would expire and the engines back in the center of the train would not pull anything, not even themselves. Heading an Eastbound into the siding there was a prescription for disaster.

It now looked like the hottest Eastbound train we ran, and this now sacred cow Westbound were going to meet there, as they always did when they were on time. This put us on the horns of a dilemma. If we held the Eastbound train at the base of the mountain, he would face a two-hour delay, and our sanity would be questioned. If we made the meeting point on top of the mountain, obeying the new instructions, anyone knowing the operation there would have no doubt that we were insane. I was glad that I was not working third shift that night.

Before he went home at about five, I again asked the chief if he was sure that we wanted to do this. He smiled and replied, "Same answer as before, the way to get rid of a bad rule is to enforce it to the letter." Now, I could see that we had brought some of this on ourselves. The hotshot Westbound was a little lighter and shorter than most trains, so we put him through the siding quite a bit, because there was less total delay that way. The shorter and lighter train could pull out of the siding and get up to normal speed more quickly. Some bean counter did not like all the ten minute delays this train incurred, so he was going to change things. I managed to issue the order of Infamy, without throwing up, before I went home at eleven o'clock.

I will not bore the reader with all the gory details, but the trains usually met about one a.m. They were about ready to leave there when the chief got to work the next morning, around seven. By this time, several other trains were stacked up behind them, causing further delay. It was a bad day at black rock, but the chief was right. The order was rescinded by noon.

We had a problem with our South line, too many cars for

the engines to pull up the nasty hill outside of the terminal on the South end, and too many cars sitting on line waiting to be picked up. We had a little meeting about it. I was the new kid on the block, and had no suggestions, but one of the dispatchers had been an agent in the area a few years back and said that there was the potential for quite a bit of business in this area if we could give them good service. He suggested we put on a road switcher that could take a thousand tons of business up the hill every morning, do whatever work there was to do, turn around and go back, making about a hundred mile round trip.

The superintendent had no problem with us trying it. We had a conductor whose wife had a business in this town, and parents living there. We knew he would take the job. This worked rather well for a while. Then this local had all the cars of his own that he could pull, so we gave him another engine.

The conductor had gone to every business we served, told them that they would they would get good service as long as the job remained, and to please give us what business they could. Soon the local had all the cars of his own that he could pull with two engines, and the crew was working ten hours a day. He now left so many North cars on line for the through train to pick up that we put on a second through train, who was taking cars up the hill for the local, and picking up the cars that he left on line for them. I guess this proves that if all you have to sell is service, it is nice if you provide it.

One Saturday morning they had to bring the snow plow into our roundhouse for some sort of service. The chief stopped at the office, and, seeing the plow in town, thought that it would be fun to take a little ride on it, and catch the first Westbound back. The traveling engineer also went with them. Being an official, he was dressed in a suit and tie, and wore coveralls over the suit to protect it and to stay warm. For some reason, the plow had to stop. The traveling engineer had to move his bowls. There was a slide fence near by, and instead of going into what usually was a very nasty toilet in the nose of the engine, he decided to get behind the fence to do his thing.

The chief took a piece of cardboard and stuck it and removed it from under the fence at the optimum times. The traveler knew that he had had a bowel movement, but now could see no evidence of it. The traveler had gone back into one of the engines, taken off and shook out every stitch of clothing that he was wearing before the chief told him what he had done. The chief thought that this was hilarious. I don't know about the traveler, but I do know he took a lot of good natured ribbing about it.

My dispatcher's phone, as most of them did, had a microphone and a speaker on it. On a damp day, if someone was talking on the chief's carrier phone, the conversation would bleed into my office. With the door closed, the offices were sound proof. This day the car distributor and the chief were talking, deciding to set a block of about twenty empty cars out at an outlaying point instead of bringing them into the yard. I called the relay operator downstairs and asked him to send a message to the train to set the cars out there, and to sign the chief's initials to it. I would imagine that this operator had been the butt of more than one joke around there, so he was happy to oblige me.

The car distributor and the chief had barely quit talking when a message came up the dumbwaiter in the chief's office. This device saved the relay operator many trips up and down the stairs. I watched the chief out of the corner of my eye. He saw the message signed by him, did a double take, and then looked at the doors to see if they were shut. He called the relay operator. "Who sent this message that you just sent me a copy of?" he asked.

"Well, I guess you did," replied the operator, "at least your initials are on it." "Come to think about it, I remember that your voice sounded different, but I thought it was the darn phone acting up again." This really puzzled the chief. With the doors closed, there should have been no way for either the other dispatcher on duty or me to hear him. He came into both offices, asking what we knew about this. I told him that this was the first that I had seen of the message. He did not believe either of us, and called the number of the phone out in his office. One could not hear it ring until he opened the door. He called the car distributor, who, of course, denied any knowledge of the message.

A couple of weeks passed. One day he was in my office, and someone decided to use his phone while he was gone. The jig was up, the condition had worsened, and you could hear the guy talking rather plainly. "You're the one!" he exclaimed, calling me several nasty names, laughing like he had pulled the prank instead of me. He knew how to take a joke, too. I teased him, telling him that I would quit doing his work for him if he called me such nasty names.

We had a tower east of town that was a very busy place. Trains from three or four railroads converged there, and, if I remember right, switch engines also to worry about. There was a jerry-rigged system of clothesline rope and pulleys used to deliver orders to trains. All one had to do was to put the orders in a holder, stick your head out of the window, attach the holder to the rope, and then pull on the rope to get the orders in a position for the trainmen to grab as they went by. The first trick operator was a woman. I commented to somebody that she was a fine operator, and how much I liked working with her. Whoever this was had a little story to tell me.

It seems that a few years back, there was a brakeman in her office when she opened the window and stuck her head out to hang the orders. This man made two very bad mistakes. The first was when he slammed the window down on her neck and raped her. The second was when he opened the window to let her loose. She had a black belt in some sort of Oriental marshal art. She then proceeded to use him as a human punching bag. He ended up with a broken jaw, teeth missing, smashed nose, broken arms, and maybe more bruises than he could count. Those injuries were minor. She then proceeded to throw him out of the window. I don't imagine that getting thrown out hurt much, but the subsequent twelve foot fall onto the near rail broke his back.

I later heard that she did not press charges until he started a law suit. I don't know how it all turned out, but I often wish that I had taken an hour or two of my time and drove over to the tower to meet her. I think that it would have been worth the trip.

Time went by. Train dispatcher positions back east were re established. There was an open job back in the Midwest that I thought my seniority would allow me to hold. A difficult decision had to be made. We liked living here, even if the cost of living was somewhat higher. I liked the people that I worked with very much, it would be hard to say goodbye to them. But we were very far from friends and family, my father was ill, and it might be several years before another job opened up in the Midwest. There was also another factor. There was a drug problem in the schools here that only seemed to be getting worse. The children were of school age. We decided to move.

CHAPTER SIX HOGBACK HOLLOW

On my first visit to the place, I overheard a conversation between two trainmen who had recently transferred there from someplace out West, Montana, if I recall correctly. One asked the other how he liked his new surroundings. The other replied, "If you like flies and horseshit, it's ideal." It was a river town, somewhat nasty in spots, but the school system was not bad, the grocery stores sold good food, the town was big enough to have whatever one wanted, not that bad a place. We found a nice home out in the country, so the town itself was never much of a factor.

I started learning my new territory from a man that was forced here, was not at all happy, and made no bones about the fact. The sooner that I could go to work on my own, so that he could leave, the better he would like it. Whenever things did not go exactly right, he was prone to answer the phone "Hogback Hollow." As the office sat in a valley, it was not bad a description of the place. I was to soon find out that there were many hog back hills to the South, causing a problem with flash floods. It did not take me long to learn the job well enough to go to work. This was the fourth time I had learned a new territory and, as with most everything, the more one does it, the easier it becomes.

The chief made me feel very welcome, maybe glad to have me instead of the unhappy guy that I replaced. I was soon to find that his heart was in the right place, but his ability to handle the officials was not nearly as good as I had become used to, but then again, who's was. He could also be indecisive, not a good trait for a train dispatcher, much less a chief. As a person, one could not find any better. I grew to like the man, considering it a privilege to count him as a friend.

The trainmaster there was somewhat of an old timer, very much liked by his men. Anybody who tried to do his job well was given every break that he could give them. He was not the type to sneak around corners to catch someone doing anything wrong. He knew how to do his job, with the least amount of effort. One morning, fusee in hand, he spied the engineer working the morning switch engine walking down the hall. : "What is this, Dick?" he asked.

The engineer, a very experienced man, replied, "Why, sir, I do believe that is a fusee."

"I do believe you're right, this was a signal test, and you passed." Now, the correct procedure to conduct a signal test would have been to hide in the weeds somewhere, and toss the lit fusee out in front of the man's engine to see what his reaction to it was. The trainmaster did not like either weeds or the mosquitoes they usually contained, and this was far easier. He was required to conduct so many signal tests each month, and I imagine that this one read "Proper recognition of a red fusee by engineer on 8:00 a.m. switch engine."

One day he called me from the Elk's club. I knew where he was, would have suspected it anyway, but the woman that tended bar there had a very distinctive laugh, and I could hear her in the background. "I need a signal test. Have the conductor on that Eastbound show a fifteen minute delay getting a red fusee at mile post 337." I told him that I could sure do that. The conductor laughed when I told him.

Now, unless one wanted to do a lot of walking over some very rough terrain, only an idiot would conduct a signal test at this location. It was morel season. It was not hard to put two and two together. "Let's see." I told the crew, "With that extra delay, what would you think of calling the outbound crew for two thirty?" There was silence for a moment, and then they replied this would be just right. The train should have arrived at two-fifteen. The conductor also showed fifteen minutes for brakes sticking. He caught on quickly.

The train rolled into town right on the call time, making

me look good. There was a nice bag of mushrooms in the back seat of my car when I got off work. With the extra fifteen minutes I had given the crew, the fifteen minutes the trainmaster had, I would imagine quite a few people got mushrooms. I don't know how many the trainmaster got, but he was able to get his signal test as part of the bargain, all without interrupting his afternoon card game. He had taken some visiting officials over the territory in an inspection car the day before, spotted the mushrooms, but did not dare to stop to pick them.

The Late Train

We had a conductor nobody seemed to like. He was a big pain in the neck. One of the older dispatchers, in particular, did not like him one bit. This was the conductor's away from home terminal. He had a thing going with the bartender at the hotel where the crews stayed. The bar closed at one o'clock. The conductor usually got into town in time to be at the hotel by midnight or slightly thereafter.

This old dispatcher decided it would be fun to delay the train enough so the conductor couldn't get to the hotel until about one thirty or so. The bartender would have gone home by then. He had everyone in on his little gag, dispatchers, yardmasters, operators, even the dispatchers of a foreign road the train ran over briefly. This dispatcher, needless to say, was sort of a rascal himself, but the conductor was married. He disapproved of this type of thing.

As far as we knew, we were able to keep this conductor celibate on our end of the railroad for quite a while. One day the trainmaster noted this train was late into the final terminal too much of the time. He asked the old dispatcher what the problem was. "It is this bone-head conductor we have. Every time he is on the job, the train gets into town too late and does not have enough time to make it. Just count the number of times he is on the job and the train gets in too late."

The trainmaster, skeptical at first, had the clerk count how many times the train was late against how many trips this conductor made. The count reveled within the last 90 days the train had been late twenty-nine times. Once due to a derailment, the remaining twenty eight when this man was the inbound conductor. With the man being on the job every third day, this removed all doubt.

The trainmaster went looking for the man with blood in his eye, and I would imagine gave him the chewing of his life. Then our strategy reversed. We gave his train every break we could. A month or so later the trainmaster again had the late arrivals counted. There were no late arrivals when the man was on the job. The trainmaster naturally assumed the chewing he gave the conductor fixed the problem; he had indeed been malingering, and gave him an even worse chewing.

Sometime after that, I had a situation where one train or the other was going to sit at a meeting point for at least a half hour. This same conductor was on the job. He was on his way home. It was Father's day and he was in a hurry. I had decided to let his train sit. It was on time. The train he had to meet was late.

The conductor saw the orders, knew where the westbound was, and knew he would get a bad delay. He came on the phone asking me to change the order. I was about to go home. I told him I was making a transfer. He could talk to the man who was relieving me; he had to live with it and could change it if he wanted to.

The man relieving me had a very acid tongue. He had been in more than one argument with this man. When the conductor asked him to change the order, he was told hell no, why would I want to change it. The conductor told him, "I think you dispatchers are all screwing me."

"We have only been doing it for about twenty years. You finally figured it out, huh?" was the reply he got. That ended the conversation.

Some Of The Boys

Not only were there some very good railroad men around Hogback Hollow, but some very unique ones. One day a section foreman wanted to know where a train would be at eleven o'clock, please ask the engineer. I thought it was unusual, but did as I was asked.

"Just a minute," I was told. Very shortly the engineer told me "At eleven I will be three poles and 34 ties west of mile post 367."

I could have fallen off my chair. I had expected an answer like "Oh, just getting close to" or "Just the other side of" whatever station. I had never heard anyone use mile posts, much less pole lengths, and surely not ties, as a reply. There was a trainman in the office getting his orders. He laughed, telling me that is probably exactly where he would be.

It proved to be the engineer had improvised some sort of a slide rule. This slide rule showed tons, horsepower, and I don't know what else. He knew how many telegraph poles were between each milepost and how many ties were between each pole. It was unusual for him to miss an arrival time anyplace by more than a minute or two. He was a real student of the game.

We had another engineer, the oldest on the roster, who had been on disability for a number of years because of a heart condition. He and his brother had a successful insurance business. Some new medication made his condition improve enough that he was able to work as an engineer again.

Most of his years had been spent on a steam engine. He was not too fond of the diesel. He could run the wheels off of an engine. Do not ask him to fix one. He showed up for work dressed in a nice suit, shirt and tie. This was one's first hint as to what mechanical work he intended to do on his engines.

About all he wanted to know was where the switch was to take an engine off line. He would use a paper towel or rag to perform this function, keeping his beautifully manicured nails from getting dirty. His shoes were always shined to perfection. Somebody nicknamed him twinkle toes. The name stuck.

The old engineer did not like his conductor very much. One day the conductor saw something wrong with the train. Instead of telling his engineer to stop, he used the emergency air valve in the caboose to stop the train. This jerked the engines enough to dump a cup of coffee in the engineer's lap.

The condition was fixed. The train took off and a short while later the conductor's voice came on the radio. "What was that for?" The engineer had jerked the caboose hard enough to throw his conductor from one end of it to the other.

"That's to show you I have an air valve up here, too," was the engineer's reply.

A short while later, the conductor again received the same treatment. This time there was anger in his voice. "What the hell was that for?" he asked.

"That is so you don't forget I have an air valve," was the answer.

There was a large hill about in the center of an area where the track had recently been rebuilt, new ties, new rail, smooth as glass. The speed limit was probably fifty-nine miles per hour.

The engineer would come down this hill at speeds approaching a hundred. One morning a wheel on the caboose broke. It derailed the caboose, turning it over on its side. The caboose slid about half way down the hill, on the rail, and came to a stop. The conductor did not receive as much as a scratch. The engineer quipped that the conductor did not wake up until it stopped, wondering how come he was no longer moving.

Another morning, the train had trouble. The conductor walked toward the head end to fix it. He managed to get this done, but for some reason the engineer was mad at him. By the time the caboose went by him it was going so fast, Spider man, on his best day, could not have caught it. The conductor was left standing there, in the middle of nowhere. I don't think he liked his engineer much either.

The engineer retired after he had his desired time in. As he put it, he retired at the moment the first wheel of the twenty-seven or so cars of grain he derailed left the rail. The speed limit was ten miles per hour. If I remember right, he was going thirty something.

This Was No Iron Horse

One day it snowed. It was the middle of April, but that did not prevent us from getting twenty-eight inches. It was my day off. The morning was cold. The sun was shining. The telephone rang. Nobody could get to work, or anyplace else, for that matter. Could I saddle a horse and try to get there?

I picked my way between the snow banks. I sat a tall horse. I am not a short man, but I could not see over the top of many of them. On the flat, where it was protected by woods, the horse plowed a little snow with his chest. We got there. I offered the use of the horse to the third trick man. The horse snorted at him, and he elected to walk home. He lived close by.

We had a crew on duty for something like twenty-three hours. The wet snow would ball up on the pin lifters, causing the train to keep breaking in two. Nobody could get to the train to relieve the crew. I think the conductor was more tired than the horse.

I have often used a little humor to make a bad situation seem less grave. This time I stuck my foot in my mouth, big time. We had a train come through town which had to pick up a switch engine. The train should have been there for five or ten minutes. The rear end of the train was blocking the BN crossing. For some reason, once the switch engine had been picked up, the engineer was unable to get the brakes to release.

The engineer was as good as they came; they nicknamed him Tom Slick, very deservedly. He was slick. But, try as he might, he could not help the situation. He changed everything he could think of and asked me if I could think of anything else.

Trying to be funny, I told him he had changed everything I could think of except the engineer. He, justifiably so, got very angry. He stayed angry for a while. I still feel bad about making the comment. I meant nothing derogatory. It was stupid on my part.

We often had trouble getting trains up the big hill. One day the sanders on an engine were not working. The train, engines at full throttle, was now going so slowly the engineer and head brakeman were walking ahead of it, sprinkling sand on the rail. I understand someone made a video tape of this. I have never seen it, but I sure would like to have a copy. They made it up the hill.

Jolted Asleep

One night, I had a shocking experience. I reached for the big metal stapler we used to staple train orders together. My chest was against the edge of a metal desk sitting on a concrete floor. The stapler came in contact with the telegraph key at the same time a jolt of lightening hit. I don't remember much of the following few minutes.

A conductor walked in, he had been a medic in WW2. He told me later I had been in shock. My heart was racing almost too fast to count the beats. I still did not feel too good when they loaded me on a stretcher. Lying down for a few moments made me feel much better.

By the time they got me out of the office after winding around the desks on the way to the back door I was feeling fine. The hallway was lined with cleaning equipment. There was a little guy carrying the heavy end of the stretcher. I am not a small man. He managed to hook his elbow on a mop handle, and almost lost me. I thought to myself it would work better if he got on the stretcher and I carried him. We would get there sooner. They kept me in the hospital over night for observation. I felt like an idiot, laying there with nothing the matter with me.

The Derailment King

Somewhere around this time I became the derailment king of Hogback Hollow. We ran a train hauling fifty cars of rolled steel once a week for about a year. At first a special train ran back with the empties. Then the empties went back on the rear end of a regular train. So far, so good.

Some bean counter saw a chance to save a little time on the cars when we handed them back to the connecting railroad. They were to be handled on the head end of the train. Bad choice. These cars had a somewhat different braking system. The brakes set and released more quickly than the standard. Every dispatcher in the office screamed in opposition to this. It was a disaster waiting to happen.

The fateful day arrived. Here came the train. I guess that the yardmaster who made up the train screamed too, but they did not listen to him. He was about to teach somebody a lesson. It was a classic example of how a train should not be made up. Steel empties on the head end, a caboose, followed by a mix of mostly lighter loads and empties. About seven thousand tons of grain were on the rear. The man may have been a sadist, too, I don't know. To top this off, there was an engineer on the train that on the homeward leg of his first trip.

I came to work at four and saw the list of the train at about five. Shortly thereafter I called the roundhouse where the wrecker was kept. I told them to get ready. They were going to the biggest damn derailment they had ever seen at about ten thirty that evening. That is only if the kid did not dump the train before he reached the hill, just before he got to Hogback Hollow.

The kid impressed me. He beat my time by ten minutes. It was not quite ten- twenty when the rear of his train plowed into the caboose on the rear of the steel empties. Seeing the approach signal yellow on the hill, indicating he might have to stop where we crossed another railroad at the bottom, the engineer went by the book. He applied the brakes in the prescribed manner.

At this point in time the brakes on the steel empties were firmly set. The grain had no hint yet that it was supposed to stop. It would not surprise me if the caboose flew forty feet in the air under the resulting impact. There was about a mile and a quarter distance between this caboose and the rear of the train, with a lot of slack in it. There were about a hundred and sixty-five cars in the train. The grain on the rear was like the energizer bunny, it kept on going and going. By the time it all came to rest, sixty-seven cars had derailed.

Cars in people's lawns, on side streets, one sat on what must have been a garage. A tank car was in somebody's flower bed.

The next morning I went out to look at the mess. I wondered, laughing to myself, if the local hardware store sold any kind of a spray to keep pests of this size out of flower gardens. Eventually the mess was cleaned up and things got back to normal.

In no way, shape, or form do I blame the young engineer. I had probably spent more time behind a throttle of an engine than he had. Someone would have had to hold a gun on me to get me behind the throttle of that train.

There are engineers who could have brought this train down the hill safely, at least most of the time. It would have been very hard on brake shoes, fuel, and the engineer's nerves. It was possible in theory. It would not have been done by the book that is for sure.

However, since they have long closed the dispatcher's office there, I shall always reign as the undefeated derailment king of Hogback Hollow, a dubious title, at best. Something like farting the loudest in church. I should not include this in my list of the most stupid bits of railroading. This one was just plain stupid, it bore no resemblance to railroading.

A Budding Friendship

About that time, the operator, who long ago, had missed giving one of my trains a train order, was again promoted. He was working in the control center. The main function of this job was to balance the engines available to each division of the railroad. The first time I talked to him on this job, he was still learning it. Another man was with him. The moment he heard my voice he quipped, "Don't worry, Joe, I don't handle train orders here."

The former operator was soon working on his own, and seemed to learn very quickly. The guy knew that I was familiar with the operation in four offices. He never hesitated to call me if he thought I could help him. Something I was more than willing to do. He was very appreciative; one seemed never to have to tell him the same thing twice. I very much enjoyed his sense of humor. At one point when we were very short of dispatchers, nothing unusual, they flew him to our office to work third trick. He had never worked the territory. He was allowed no time to learn it, going to work with a timetable in one hand, a pen in the other. I came to work the next morning and met him face to face for the first time. I expected to find him in a frazzle and the railroad in a mess.

To my surprise, I found him looking as fresh as a daisy, laughing about this or that. There were some rough edges. Maybe some of the trains were not lined up to meet in quite the right place. A line up figure might be a little far off the mark. But in general, things were in good condition. A very remarkable feat, I thought. He was with us for only a couple of days, but it was enough to impress me. Over the next year or two we averaged talking three or four times a week, depending on what shifts we were both working. A friendship blossomed.

CHAPTER SEVEN THE LUMP-LUMP DAYS

From Good To Bad

Then misfortune struck. Our trainmaster had been promoted, very deservedly so, and we lost him. To top it off, Willy lump-lump took his place. We went from having an official the crews liked, respected, and cooperated with, to one they all soon would hate. We could now expect no cooperation from anybody, justifiably so. Lump- lump criticized everybody for most anything every chance he got. His favorite targets were the train dispatchers

I was the union representative for the office, and I had to deal with him. The chief tried to talk some sense into him. He was not able to dissuade him. More often than not Lump-lump screamed at him too.

To put things very simply, we had war. One of his pet peeves was the delay trains had at meeting points. I said to him his former chief had told him that he wasn't able to get two wheel barrows of shit around each other in a forty-acre cow pasture without tipping one over and delaying the other one. He turned red in the face and denied it.

I looked at my watch and said "The man should be working right now; let's call him up and ask him."

Lump-lump replies, "I don't want to talk to that son of a bitch."

"I bet not, I wouldn't want to either, if I were you. All he would do is eat your ass out again. I think he is pretty good at that."

"Well," he said, "maybe I wasn't any great shakes as a train dispatcher."

"They make umpires out of baseball players who can no longer see well enough to play ball, so why not officials out of dispatchers who can't dispatch trains," I needled him.

"You're getting awful close to insubordination," he told me.

"Well, you maybe are right," I replied, "But see if you can't lay off of the guys a little bit."

This helped for about a week, but he soon was back at it again. He got in a row with one of the road masters. It came to blows. The road master, a huge man, gave him a swat that sent him over the side of a bank. Willy lump-lump now sported a pair of very large dark sunglasses, and did not show up much.

About the time the sunglasses came off he got into another row. This time he fought with the assistant road master in another territory. This guy was not very big, but at one time he was one of the best middle weight boxers the Marine Corps had. The man was still as quick as a cat. I will leave it to the reader's imagination to figure out the result.

Willy lump-lump now had been told one more fight and he was gone. To seek employment elsewhere, this was his last chance. One day I came to work on second trick. He was berating the first trick man for something I felt was not at all his fault. Words got very heated.

I told him "I hear things are got going too well with you and your boss. Moonlighting as a prize fighter isn't working too well for you either. You can try hitting me, if you need the practice." For a moment, I thought he was going to take a swing at me. He managed to only turn red and stomp off to his office. I don't think he liked me very much

A Definite Improvement

Then something good happened. The superintendent who had the nervous break down came back to work as the chief rules examiner. He had undergone a huge personality change. He now always seemed to be in a very good mood, laughing and telling jokes, something he had never done. He was now awarded the Superintendent's job encompassing my territory. He hated Willy lump-lump with a passion, and now would be his boss. Railroading might get to be fun again.

In a week or two the superintendent called about something. Willy was out in the country someplace and came in on the dispatcher's phone. With the boss listening, I thought I should be on my best behavior, and very courteously told him, "One moment, please, sir, I am on another phone."

"Let the son of a bitch wait," said the superintendent.

Before long, another "Dispatcher" came over the phone, this time a little more insistent I ignored him. A short time later another "Dispatcher" came over the phone, this time with a real bite. I pushed the peddle on my phone down briefly, to let Lump-lump know I was still talking. By now the Superintendent had finished his business and was talking about the weather, last week's ballgame, whatever. I knew what he was trying to do, piss Willy off by making him wait.

I was getting behind in my work and told the superintendent, "Well, I had better go see what lump-lump wants, he is standing out in the rain at a call box."

"Good," replied the superintendent, "I hope the son of a bitch drowns."

Willy screams into the phone, "Tell whoever you are talking to, to shove the phone up their ass and talk to me."

"I'm talking to the superintendent," I informed him.

"Please tell the superintendent to take all the time he needs," he replied.

Trying to muster up the sweetest voice I could, I told him, "Oh, sir, that is not necessary. He is on the speaker phone and can hear every word you say."

There was dead silence on my dispatcher's phone. Someone stuck his head into my office to see what all the laughter was about. I don't know which one of us laughed louder, but the superintendent had achieved his goal.

It took Willy about half an hour to muster up the courage to call in again. I very politely informed him I had gotten so far behind in my work, talking to him and the superintendent; I could not talk right now. My trains would be standing, please call back in a half hour. He did not argue. Well, the boss told me to make him wait. I did.

We dispatched a small subdivision, jointly owned by three or four railroads, including our own. Two of the foreign roads ran a daily local each way. Years ago the locals had been shown in the timetable as third class trains. This turned out to be a rather awkward way of running things. The chief rules examiner issued instructions for them to be run as extras. I, as well as the other dispatchers, thought this was a good idea, it made our work easier.

It was time for a new timetable to be issued. Lump-lump had seen an old timetable somewhere and decided to put the schedules back into the new one. We had some heated arguments about it. They ended when Lump-lump told me that the schedules would be put back in.

I told him "I bet you I can annual just as many schedules as you can put in there." Needless to say, when the new timetable came out, it sported the new schedules.

I promptly issued a train order annulling all of them for a week. When he saw this, he started screaming at me, red faced, his arms waiving like a windmill.

"Don't scream at me," I told him, "I am just following instructions. If that is what the chief rules examiner wants done, that is what I am going to do."

"Fuck the chief rules examiner," he screamed, "I am telling you I want those schedules to stand, and that's what you are going to do."

"Afraid not, not in this lifetime, anyway," I replied.

"You mean you are threatening to annual them again next week?" he stormed back at me.

"No, no threat," I replied. "How clear can I make myself? I have told you over and over again what our instructions are, and am telling you again that we will comply with them." "No matter what you say, period."

Now the man was really angry. "I am taking it up with the Superintendent," He stormed, and "Maybe he will fire your ass."

I hesitated, like I was thinking about it real hard. "Hey, that isn't a bad idea. At least it should settle the argument." What I did not tell him was these instructions were issued over the signature of the superintendent when he was the chief rules examiner. He would find out soon enough. If he wanted to walk into a door, let him. He never mentioned the subject again. When the next new timetable came out, the schedules were gone.

A train on the branch line was doing some kind of work requiring two or three hours to complete. There were no other trains involved so the train dispatcher ahead of me, wisely, gave the train exclusive use of the track. This freed up the man who otherwise would have to act as a flagman for other chores.

Lump-lump was conducting signal tests. He asked me for a copy of the set of train orders the train had. He took off. Shortly I heard him screaming at the conductor because he did not have any flag man out to protect the rear of the train. Not seeing any order specifically issued to relieve the crew of this chore, he got angry. His wits flying out the window with his temper.

The engineer on the job quickly came to his conductor's defense. "Hey, we have right over everything out here; we don't have to protect ourselves against any other trains in any direction. What should we be flagging against, trainmasters?"

Lump-lump, now realizing he was wrong, quickly backed off. I called the engineer. "Please stop by the office when it's handy for you," I told him "I have a story to tell you about a couple wheel barrows full of manure." I could tell that the engineer was puzzled by the request. He assured me he would stop. The trainmaster, I am sure, knew exactly which two wheel barrows I was talking about. I could not resist needling him a little.

One dark night Lump-lump was at it again, more signal tests. This time he was in an area with double track. The Westbound main veered off to serve a rather large town. The Eastbound main having been laid a couple miles away to avoid having to go up a rather large, steep hill.

Lump-lump wanted to signal test a Westbound train, and asked the young dispatcher on duty where this train was. After being told, Lump-lump set off in the rain to position himself to do his test. The dispatcher may have been young, but he had not fallen off any turnip truck. He figured out what Lump-lump was going to do.

Forty or so minutes later Lump-lump came back on the dispatcher's phone, demanding to know where the train was. "Oh," said the dispatcher, "he went by you about twenty minutes or so ago."

"Like hell," said Lump-lump, "I was standing close to the track. I never saw him."

"How could you, he would have been about two miles from you," replied the dispatcher, acting very innocent.

"You mean you wrong tracked him?" screamed Lump-lump, "And let me stand here out in the rain?"

"You never asked me which track he would be on. Did you get wet?" he asked, again acting as innocent as a new born baby. Aside from the profanity, Lump-lump did not say much. I imagine this dispatcher still laughs every time he thinks of the incident. Such can be the life of a railroad official that mistreats his men.

This young dispatcher rose through the ranks very quickly. He became the chief rules examiner, but he loved to run an engine. At one point they went back to doing exactly that. The last I heard of him, he was with Amtrak. I helped train a lot of young dispatchers. This one learned the quickest of them all.

He is one sharp fellow.

One thing our chief could do better than anyone else was find good candidates to become dispatchers. None of them were local products. I have no idea how he went about finding all of them, but find them he did. We had a string of them pass through the office, each as sharp as the last.

All the railroads had eliminated enough operators' positions to make it hard to find dispatcher candidates. I think that about every railroad in the United States was short of dispatchers. An experienced dispatcher was almost impossible to hire. If an experienced dispatcher wanted a different job he could provide references and apply. He would only be asked when they should have the moving van in his driveway.

Most of our trainees did not like Hogback Hollow too well, Lump-lump less, and had little incentive to stay. At one time, shortly before I retired, I think that the BN had six chief train dispatchers. I had helped train four of them.

When I became the working chief I hired one dispatcher. He was a local boy, working a job about two hundred miles away. I liked the fact he seemed very conscientious, but never seemed to get too shook up about anything. He seemed very sharp, willing and cooperative.

I called him at home. His wife answered, he was out working in the garden. I should have told her what I wanted, but did not think of it. Chief dispatchers do not ordinarily call operators at home on a social basis. He thought he had done something wrong. He did not know what, but was very apprehensive when he answered the telephone. He was much relieved when I told him I wanted to promote him. I felt bad about not telling his wife why I had called.

He agreed to give it a try. We soon had one of the best students we could ask for. He learned quickly and well, and never complained. If my memory serves me correctly, he once worked about three months in a row without a single day off.

He seemed to be in as good a mood the last day as he was the first.

Whatever high expectations I had for him, he exceeded by far. The only reason I could see that the chief had by passed him was his youth and slight lack of experience. I tell him I still do not know if I did him a favor or a disservice the day I called him.

I told all the young train dispatchers that passed through Hogback Hollow, laughingly, that if they could work in an office Lump-lump frequented, they could work anywhere. There was probably more truth in my words than I realized.

Branch Line Shenanigans

We dispatched a branch line that did not amount to much, with the exception of one huge plant located along it. I would imagine it did a million or two dollars worth of business with us each month. It was sort of a ragtag operation. There was no way to put up a decent meeting point, or to change one, without a lot of luck.

An old conductor suggested we might solve our problem by putting up a meeting point at the longest siding, have the conductor call in when he was ready to leave the plant. The outbound crew at our terminal would be called for duty an hour and a half later. It sounded good to me, so we tried it.

His suggestion worked out well for the most part. Instead of having to go to work at four thirty, the standard call time, the crews often did not have to come to work until six or seven in the morning. This gave them extra time at home. They all liked this, and busted their butts to make it work. The dispatchers liked it, one less meeting point to worry about.

If the crew could not make it into the terminal within the allotted twelve hours, they always managed to get the train within twenty-five miles of either terminal. The maximum distance a switch crew could come and get the train. It might cost an hour or two overtime, but it was much cheaper than having to call another crew and pay them a day's pay.

Lump-lump caught wise to what we were doing. I was working third trick at the time. He asked me how come we were doing things this way. I told him it was for everyone's benefit. The dispatchers and crews liked it, and the plant seemed to be getting good service.

He blew up at me. "I am not running this railroad for the convenience of the train dispatchers or the God damned crews." He promptly stomped off to the yard office where the clerk that called the crews worked. Instructions were issued to them to call the train crew at four-thirty, if they were rested, no excuses or exceptions.

Call them they did. I tried my best to give the trains some kind of decent meeting point, but it seldom worked. The crews did not like this change one bit. They had various little ways to let Lump-lump know this.

One of their favorites was to stop one station short of the meeting point with the other train already there. They would go to eat for an hour, which they were allowed to do. When they got to the meeting point the crew on the waiting train that had maybe been sitting there for two hours or more would decide to go eat. This delayed the other train for an hour.

Instead of getting the trains within the twenty five-mile limit, they would always manage to leave them far enough out so we had to call a relief crew. We now had to use four crews a day instead of two and spent a lot of money on taxis. All this without accomplishing one bit more. One thing the crews always managed to do was give the plant excellent service. They had no intention of biting the hand that fed them. This I was glad to see.

A lot of comments were made around the office about what the superintendent would say when he saw how much money we were spending here. "Someone's head is going to roll," or

"The shit will really hit the fan." All within all within lump-lump's hearing distance, and loud enough so he would be sure to hear them. Anything to brighten up his day. It did not take long for him to rescind the order. Things went back to normal.

They did not stay normal very long. Lump-lump was mad as hell at the crews. He caught both of them in the depot at the meeting point one day, eating a sandwich and playing cards. If he wanted to save the railroad money and a lot of people, mostly himself, a lot of grief, he would have bought them each a Danish. He could have sat down to play a hand or two with them and told them what a good job they were doing, then got out of their way. Such was not to be. He suspended both crews without pay for ninety days for playing cards on duty. This time he had really fouled his own nest.

We ended up with two entirely new train crews. I imagine they all had some experience around the plant, but most likely had experienced men with them. Nothing seemed to go very well. We again had to use four crews instead of only two, and sometimes did not have enough men. The men we had out there were very unhappy with what happened to their buddies. They knew they would be out there twelve hours, so why hurry. The outbound at our terminal would sit for lack of a crew. Things were in a real mess and getting worse. The plant was not getting the service it wanted and deserved anymore.

One day the phone rang. It was the plant manager, most unhappy. "What happened to all the good service I have been getting?" he asked. "I called the trainmaster's office and complained, but things are getting worse instead of better. What happened to all the regular men, I never seem to see the same man twice."

"Yes, I know," I replied, "I am not surprised. Someone told Lump-lump, years ago, he could not get two wheel barrows full of manure around each other in a forty-acre cow pasture without delaying one and upsetting the other. He probably was right"

I gave the plant manager the whole story and suggested, "Why don't you call the division superintendent? He is probably the best operating officer this railroad has, and I am sure that he can help you. The obvious solution is to put the suspended crews back to work, but I am just a little old train dispatcher, and I don't have any authority. I sure am sorry for all of the grief this is causing you, though. I can give you both the superintendent's office and home phone numbers, if you want them and you can quote me, if you want to." He took the phone numbers, thanked me, and hung up.

For the life of me, I cannot remember what transpired after that. I think they did put the crews back to work early, but had trouble rounding them up. A couple of them were in Florida, and another was in Acapulco.

Lump-Lump Plays In The Snow

It was winter; we had snow, lots of it. Adams-apple deep to a tall giraffe in some of the cuts. We had a derailment, a rather bad one. Lump-lump was at the site. It was bitter cold with a wind blowing. He had gone about thirty six hours without sleep.

The derailment was cleared up and we could run trains again. Lump-lump came in on my phone. "Be sure to get that first West guy called," He instructed me.

"The crew is here, putting the train together. As soon as they get that done and the air pumped up, I'll try to get him out of town. That might take a while, though, cold as it is, and those cars have been sitting for a long time." He took no exception to this.

"How about the next West guy, is he called?" Lump-lump asked.

"No, out of crews" I replied

"Is the connection called for the guy who was derailed?"

He now asked.

"I told you. I'm out of crews."

This time he cussed a little. "Then at least get the branch called and get some of the cars out of the yard," was his next pipe dream.

"Let me tell you a story," I told him. "This woman was going to have a birthday party for her five-year-old boy, and went to the store to buy ice cream. Please give me a gallon of vanilla, a gallon of strawberry and a gallon of chocolate," she tells the clerk.

"Lady," says the clerk, "I have vanilla, I have strawberry, but no chocolate."

"Then give me a gallon of vanilla, a gallon and a half of strawberry, and a half gallon of chocolate," she tells him.

"Lady," He replies, "I told you I have vanilla, strawberry, but no chocolate."

"Then give me a gallon and a quart of vanilla, a gallon and a half of strawberry, and only a quart of chocolate," she says.

"Lady, how do you spell the van in vanilla?" he asks.

"Why, v a n," She replies.

"And how do you spell the straw in strawberry."

"Why, that is easy, s t r aw," She tells him with an inquiring look.

"And, finally, lady, how do you spell the fuck in chocolate?" he asks.

Shocked, the lady replies, "There is no fuck in chocolate."

"Now you get the idea, Lady, now you get the idea." He tells her.

"Well," I tell him, "That guy was out of chocolate, I am out of crews, and now do you get the idea?"

Lump-lump did not seem to think my little joke was very funny and started quizzing me on what happened to all of them. "Well," I told him, "About half are at the other end of the railroad, like they usually are. We had to run a plow East this morning, dog-catch the branch line crew, one crew is on the train that derailed, two are on the ones behind him, one is pumping air here, some are getting their rest, and the caller told me some idiot fired the rest of them." This time my fish was evidently too tired to rise to the bait. He got off the phone.

It took him about an hour to get to the office. He was unshaven, hair all messed up, eyes bloodshot with dark circles under them, and very wind burned. He looked like he was dead on his feet. Wishing to further brighten his day I told him, "You look like shit." He sat down in the extra chair in the office with a big sigh.

"And that is about how I feel," he managed to reply.

For the first time since I met the man, he seemed willing to talk without getting angry, why I do not know. Maybe the full effect of what he had done to himself had a chance to soak in, or he was just too wore slick to get angry. At any rate, I thought this might be a good time to try to talk to him.

"You know," I started in on him. "In all of my years on the railroad, I have never met anyone who insists on making life as miserable for himself as you do. For instance those eight guys you suspended were probably some of the best we have. They all have the comp. Most of them are someplace where it is nice and warm, getting all of the sleep they want, enjoying life and laughing their asses off at you."

"In the meantime, you are mucking around in the snow, re railing cars, not getting any sleep, and freezing your ass off. We now use sixteen men a day instead of eight on the branch, and the work is not getting done as well as before. Our trains are sitting because we are short on men. I know why you fired them, you were mad at them."

"It seems like that every time you get mad you almost invariably do something stupid. Everyone suffers for it, you most of all. Now, why don't you go home and get some sleep before you keel over, the place will be here when you get back.

And, incidentally, for whatever it is worth, I think you did a fine job getting that derailment cleaned up."

The man looked at me, nodded, and left without saying a word. What I had told him was true. The crews all had income protection insurance that paid them if they were suspended. The compliment about getting the derailment cleaned up was also true. He had done a nice job. I thought maybe if he found out how much better a compliment feels than an ass-chewing, he would maybe start handing out a few, instead of a constant flow of criticism. Something I said must have reached him. I never did hear him compliment anybody, but we now had a truce. An uneasy one, but a truce none the less, far better than what we had before.

The Run Away Train

Time wore on. I was working third trick one morning. A car with a defective wheel split the switch where the double track turned into single track, at the extreme west end of the territory. Several cars were derailed, blocking both main tracks. There was a Rock Island freight train parked on the Westbound main behind the derailment and a local several miles behind him, working on the same track.

The conductor of the local came in on the phone, saying he was getting short on time, had to either make a run for the terminal or be relieved on line. I told him about the derailment and to keep working. I would send a cab out to get the crew. He should put the train into the clear on the siding before he went home.

Time passed. The conductor came back to the phone informing me he would not be able to get his train into the clear. There was a car with a rigid truck that rode up on the rail when they tried to back into the siding. They would derail it if they persisted. I told him to tie the train down so it could not go anyplace, leave the cars and the engine on the main track and go home.

About forty minutes later, my dispatcher's phone went out. A while later the operator at this location called me on the city phone, all excited. He told me there had been a horrible accident. The local plowed into the rear of the Rock Island Freight train. The local hit the caboose so hard it drove it into the box car full of corn just ahead of it.

The conductor and rear brakeman were in the caboose. One received a broken arm, the other a broken back. They almost smothered in the corn, one pulling the other out. How this feat was accomplished I will never know.

The engine and all eight cars remained on the rail. There was three to four feet of corn on the running boards of the engine. There is no way anyone could have gotten out of it after the impact. The brakes were fully applied. It should have been impossible for this engine to have moved even one inch.

The grand-daddy of all investigations was held. A typical investigation lasts maybe a couple of hours. This one lasted three days. I have to hand it to Lump-lump; he knew how to conduct an investigation. I don't know how many of them I attended over the years as a witness, an office chairman, or a general chairman, but it was a lot of them. Never have I ever seen one orchestrated more smoothly than this one. The man never missed a beat in the entire three days. I do not think an experienced trial judge could have done better. Maybe Lump-lump had missed his calling.

Outside of the number of witnesses and its length, there was only one novel thing about this investigation. The engineer on the local was very inexperienced. The union representative for the engineer had a personal vendetta regarding the length of training a new engineer received. I am sure he had a very valid point to make, but this was not the time or place. He was, in effect, hanging his own man. I could have stuck an apple in his mouth. Lump-lump cut him off more quickly and effectively than an apple would have.

Witnesses we had, people from Westinghouse, who had manufactured the brake system and General Motors, who had manufactured the engine, roundhouse and car foremen. The list goes on. I did not see the pope there, but the huge room was crowded, and I could have missed him. I think there were more people there than at the Nuremberg trials. Both the Westinghouse and General Motors engineers testified that it should have been impossible for this train to move.

As I had been on duty when the incident occurred, I was there as a witness. As I recall, I was gone five days, two for travel, and three at the investigation. I do not think I spent more than two minutes on the stand. There was not much I could tell them.

The investigation was held at Lump-lump's spa. I never saw the bill for the room. By looking at the menu in the dining room, I could understand why the superintendent was upset with the man's expense account. Figuring my weeks wages on top of my expenses, my two minutes of testimony were pretty expensive. I felt like a pretty pampered dude by the time I went home.

Later, I got a copy of the transcript of the investigation by mail. It was a stack of 8x11 single-spaced papers that must have been close to four inches thick. The only possible use I could have had for it was as a door stop. I had one of them.

I heard the expense for the stenographic service alone was over nineteen thousand dollars, a lot of money in the 1970's. The crew involved drew some time off without pay. Being it was hot at the time; I imagine they went to Canada to enjoy their rocking chair pay. If so, I hope they caught a lot of fish. I never did hear what was said when the bill for all this hit the superintendent's desk.

I did not dwell on it, but idly wondered for many years exactly what had happened. I had been retired for about ten years when one day an old western movie was on TV. I was not watching it, but I heard an old railroad expression used. A light came on. I now thought I knew what had happened. A one in a million fluke maybe, but an explanation. I never did blame the engineer, and if what I now suspect is true, it would exonerate him completely. I could never prove my theory in a court of law, nor would I want to, but I can go to my grave feeling as though I know the answer, or at least a possible explanation

Things went along rather uneventfully for quite some period of time. Until one evening we had a derailment, a minor one. A broken wheel caused one end of a car to derail, not far from the head end of the train. The proper calls were made. The car department and Lump-lump were on the way when the yardmaster came into my office with a list of the train. There was a block of maybe fifteen cars, just ahead of the derailed one. They had not been switched at the terminal. They were just slapped on the head end of the train to fill him up closer to tonnage limits. By now, the crew, the engines and the head end of the train were on their way to the next station. They would get rid of the head end so they could go about re railing the derailed car.

"How do you think it would work, if the crew is willing, to have them switch the head end out when they get there?" asked the yardmaster. "By the time the train gets here, we will be classifying a Westbound and the industry job will be up town. I won't have an engine to do the switching with, and it would save two to two and a half hours delay to the train, if the crew is willing to do it."

"How long do you think it would take the crew, half an hour maybe?" I asked him.

"Maybe or even a hair less." The yardmaster replied.

"Well, he should be back to derailment with his engines about the same time that the car department should be there.

Let's do it, if the crew is willing." I responded.

When asked, the crew was more than willing, saying they would rather have something to do than just wait. Time goes by, and Lump-lump comes on the phone. He demanded to know where the engines were, sounding very unhappy.

"They are switching the head end up," I told him, and "They should be back in a few minutes." The man let loose with a tirade I have not seen surpassed before or since. I promptly turned the volume on my speaker down to where I could barely hear him, put my feet on the desk, and started playing my imaginary violin for the yardmaster's amusement. He was there, wondering what time the engines would be back. His phone rang, and I could hear him laughing clear back to his desk.

Typical Lump-lump, I thought to myself. People are doing him a favor, and he is standing somewhere on the edge of a bean field, ranting and raving about it. "Well, the headlight is showing up," He said. Silence "I said the headlight is showing up," he said in a louder voice. Silence.

"I said the head light is showing up," he now screamed. Silence. "Dispatcher," he said, repeating the word in an ever louder voice. "Fuck it," he finally said.

I could care less if the headlight was showing or not. It would take the car department a good half hour to jack the car up, change the broken wheel, and set the frogs. I was not about to talk to him. As far as he knew the telephone went out, he had been chewing out the nearest post.

In due time the car was rerailed. Lump-lump was calling me on the radio, evidently in his car, on the way back to his motel room. I let him call me a couple of times. Then I called the engineer of the train and asked him to please tell the trainmaster, if it was indeed he that was calling me, I could not hear him, and to speak louder.

I could probably have heard him without a radio if I had

opened my office window. I told the engineer I still could not hear him. All of a sudden, the engineer could not hear him either, and called his caboose to see if they heard him. "Well, I did earlier, but I don't now, all I hear is a lot of noise." Lump-lump started calling the car department and the assistant road master. For some reason, they could not hear him, either. Everybody out there took turns asking each other if they could hear him. Nobody could. Judging from the amount of chuckling in the background, someone had over heard his tirade. The word had spread. They all agreed there was a lot of noise on the radio that night.

A while later, the city phone rang, it was the superintendent. "How are things going at the derailment?" He asked.

"Good as could be expected," I replied. "They have the train back together, and should be leaving. The only Westbound that I have out there should be in the hole, waiting for him, by the time he gets there. We got about a three-hour delay out of it, but the crew should make it in ok on their twelve hours. The only thing that went wrong was I got the ass-eating of my life from Lump-lump."

"Why," he asked.

I went on to explain exactly what we had done and why. "That son of a bitch," he said, "Just wait, I'll fix that."

Two or three days later, I was just getting off duty when the Superintendent walked in, Lump-lump in tow. There were several people in the office already, two dispatchers, the yardmaster, and I don't remember who else. "Say," said the superintendent, "I just stopped by to tell you fellows what a good job you did with that derailment here the other night. Smartest damn job of railroading I have seen in twenty years."

And he went on from there. Lump-lump stood there, suddenly finding his shoes to be very interesting. Every time he looked up, the superintendent would give him a look that would have frozen molten lava. When the superintendent seemed to be running out of steam, I would make some comment on how good the yardmaster's ideas always were, how helpful he was, polishing his halo the best I could. What I said was also very true.

The superintendent left the office with his arm on the yardmasters shoulder, as if he intended to bestow knighthood on him at any moment. This was to be the last time that I ever saw the superintendent. Some time later he went on vacation, and retired without going back to work. I guess he did not like goodbyes.

Lump-lump went back to his office acting like a whipped puppy. I am sure he heard the laughter break out before he got his office door closed. Everybody there knew the story. They could no longer contain themselves. The superintendent had made his point without using one four-letter word. I am sure the humiliation Lump-lump must have felt was far worse than any chewing he ever got.

One would have thought that Lump-lump should have learned something from all this, but such was not the case. He was now worse than he had ever been. His chief targets now seemed to be the train dispatchers. He second-guessed them and chewed on them every chance he got. I warned him that he had stepped one tope over the line. We were not going to put up with his happy horseshit any more. If he wanted to make all the decisions he could make them, not second-guess them. His phone started to ring, day, night, or anywhere in between, always with a problem for him to solve.

Lump-lump complained to me that he was not getting any sleep. I had been working the swing job, two first tricks, two seconds, and one third. I told him if he kept complaining I would go on third trick, and then things would really get bad.

One of the dispatchers started bringing an alarm clock to work, setting it at two hour intervals so as to not forget about calling him. If a traction motor had to be cut out of an engine, his phone rang. It got so if a cow shit on the track, his phone rang before the echo of the splat came back.

I came to work on third trick. He had not left any tie-up. I imagine he remembered my threat to go on third trick. He must have known it was my night to work. Second trick had last talked to him at about ten o'clock. He was out on the line someplace.

The first thing I did was call the number of the spa and ask if he was there. He was there, but left word that he was not to be disturbed. As luck would have it, about two-thirty we had a derailment. I tried calling the spa, same answer. I made the arrangements to get the necessary help on the way, and called the superintendent. "Sorry to disturb you but I have a problem. We had a little derailment; I have all the necessary arrangements made to pick it up. Lump-lump is at his spa and the desk says he left instructions not to be disturbed. I don't see where I need him for anything. I know that an official is supposed to be there and am wondering what you want me to do."

For the first time in recorded history, a superintendent who had just been advised of a derailment, failed to ask who, where, and how bad. "I'll fix that, call you right back," is all he said. It did not take five minutes for the phone to ring. "Well, they told me the same thing at that damn spa, but the vice presidents of operations and transportation are in town. I woke 'em up, and they are taking a cab to go get him. Now, what have we got?" I told him in brief what was involved. "Thanks for calling me" was all he said, but it was another first.

It was about a twelve-mile cab ride from down town to the spa. There were probably a dozen people at this location he could have called, including a full superintendent. Yet he called two vice-presidents to do the simple chore. I do think he had a little more in mind than a simple wake up call. If I were a vice president, I would not appreciate a call like this. I imagine they did not either. I thought to myself I may not have any high-class call girls in my world, but I sure do have some high-class call boys.

When the derailment had been taken care of Lump-lump came in and reported the fact in a very courteous voice. I thanked him the same way and let him sleep the rest of the night. That is, if he could.

A few days later, he asked me to stop into his office when I got off work. We had rescinded our wake-up program; it had served its purpose. I found him sitting with his elbows on his knees, face in his hands, looking dejected indeed. "The harder I try, the worse it gets," he told me.

"It doesn't have to" I told him. "You just have to decide what you want. If you want us to make you look good, we can do that. If you want us to make you look bad, we can do that too. I would rather, if given a choice, try to make you look good, but the choice is yours." All I got out of him was a half smile and a slight nod of the head. This was the end of the war.

He spent about a year there after that, and I don't think there was ever a single case of him being abusive to a train dispatcher again. When he left, he was more so a friend than an enemy. He ended up working for another railroad, first as an assistant superintendent, then as superintendent. There was nothing wrong with him as an official, if he kept his temper.

CHAPTER EIGHT A BETTER ERA

A Great Boss Arrives

The operator who had missed the train order and was now working at the control center called. "Your ass is grass," he gleefully told me. "I'll be down that way next week to look for a house. They promoted me; I am now a trainmaster." Somehow, the promotion did not surprise me.

I don't remember exactly how long the new trainmaster was there. It was not long enough. He backed the train dispatcher's decisions, trying to make things easier for them. He was a total joy to work for. He had a routine. Most mornings he would come in and look at the train sheet. He would then tell me "I'd fire you, if I didn't shudder to think who they would replace you with," slap me on the back, and laugh all the way back to his office. I was working third trick at the time, so I heard this many times.

When Lump-lump was there, the men would come back from a derailment tired, sullen and silent. Now they came back probably as or more tired, but they sounded like a bunch of guys on the way back from a night on the town. You could usually hear them before you could see them. The trainmaster's infectious laugh rising above the rest, not seeming to care they were tired. He made sure they had food when they needed it. He would often be as dirty as they were from working right along a side them.

Lump-lump would cut every dime of overtime he could; this guy would try to pad it a little. He got far more for his money than Lump-lump did.

Over time, the trainmaster and I became better and better friends, both at and away from work. It was a good period of time; we worked hard and had our fun. There were only two negative things about it. We knew it would not last. This man would go upstairs quickly. The other was that we were having too many derailments, bad ones.

A lack of track maintenance was part of the reason for the derailments. The other was the now common use of the jumbo hopper car. They had a high center of gravity. I do not think they were designed for the type of railroad we ran them on. They were too heavy for the type of rail we had, rocked like a drunken hobby horse, wallowing over the railroad with the grace of a file of pregnant hippos.

An Unwanted Promotion

Somebody decided, instead of fixing our track, we would run over the Rock Island for the majority of the distance involved. A chief someplace else retired. Our chief took his place. We were now to have just a working chief. Having to dispatch only about a hundred miles of railroad and a branch line, the work load was much less, but nobody in the office wanted the job. They did not want to deal with the bean counters. I do not blame them. Still being the youngest regularly assigned man in the office, I was forced on the job.

Now I wore two hats, neither of which fit very well. I was both the boss and the union representative. I tried to get somebody to take the union job, but nobody would. If there was ever a conflict of interest situation, this was it.

I complained to one of the guys about it. He suggested I set up a small table with a chair on each side of it, and shift from chair to chair, to present both sides of any issue. The conflict of interests issue never became a problem. I was not comfortable with the "mother hen" role, I felt like my tail feathers were too short.

We Get Rooked On The Rock

Running over the Rock Island proved to be a disaster. The Rock Island had a new chief operating officer who, as near as I could tell, knew very little about getting trains over a railroad.

He had this pet train he refused to see delayed, crucifying his whole operation in the process. His railroad got so full of trains sitting on sidings that not much of anything moved. I put a crew on duty to relieve a crew sitting on a siding. They got over there in a cab and sat until their twelve hours expired, never turning a wheel.

We had a new trainmaster, an ok guy, but there was little he could do. His hands were tied. Then, slowly, the trains started to run over our track again. Having fewer of our trains over on their track did not seem to help a great deal. The operation still stunk. Now all of the trains were back on our track, and more besides. Without a full time chief to help, we were more than busy.

Then the Rock Island operating officer came out with a new edict. We had always run over a short segment of Rock Island track on the east end. Now, no train was to be let out ahead of the sacred cow unless it was at least an hour ahead of it.

We had a train that seemed to get an hour or so of unnecessary delay following this train almost every evening. I left a note for the third trick man that if this happened again, to wrong track our train around his on the double track at the west end, giving our train the faster track. Then run our train into the terminal ahead of him. I also arranged for the yardmaster to leave the rear end of our train hang out of the yard for at least an hour, effectively blocking the sacred cow. The plan worked to perfection.

The next morning, the shit hit the fan. The Rock Island operating man called me, swearing, and I hung up on him. I got several phone calls on the subject, wanting an explanation for all the delay to the sacred cow. "Simple," I would tell them. "If he screws us, we screw him, he gives us an even break, we give him one." To my pleasant surprise, I heard no cuss words, but I did hear a few chuckles.

Later that day, the phone rings and the caller identified

himself as senator so and so. I had more than a sneaking suspicion of what he wanted. "Just a moment, please, sir," I told him as politely as I could. Then I put out a couple of train orders. I thought that making him wait would give him the impression that he could not shove me around. Plus, it gave me a moment to gather my thoughts. "What can I do for you sir?" I asked him as politely as I could.

"I had this call from a gentleman on the Rock Island railroad, he seems to be very unhappy with the train operation down there," he said in a polite voice.

"So am I Sir, he seems to delay one of our trains needlessly almost every evening," I replied.

The senator seemed unsure of what to say next, but came up with "I just want you to know the operation of this train is important to the Rock Island, and the operation of the Rock Island is very important to the state."

"I would think the operation of this railroad is also important to the state," I replied, "and besides, they are the ones who sign my paycheck."

"And," replied the senator, "The man said you hung up on him."

"You sir, seem like a perfect gentleman," I replied, "And I am sure you do not relish being swore at, and neither do I."

The senator asked, "But what can I tell the man about the operation of his train? Are you threatening to do the same thing again?"

I replied "Sir, I do not believe in making threats. You can tell the man that, under the same set of circumstances, it will happen again as surely as dawn follows the dark."

"Well, ok, thank you," he said.

"Thank you too, sir, for your interest in the matter. It was nice talking to you," I replied. The conversation ended. I had been so nice to the guy I had wanted to puke. I did not think that we needed the government for an enemy.

A couple hours later the phone rang. It was the Rock Island operating officer. This time he did not swear at me, but instead said, "I should hire somebody to come down there and shoot you."

In as kind and polite a voice as I could muster, I told him, "Sir, I won't worry about it too much, if you handle that the way you handle your train operation, they surely will miss." He slammed the receiver down in my ear, but our trains no longer got delayed. I never talked to the man again, and that was soon enough.

The Gentle Giant

As if we did not have enough problems, we now had a group of boys come in on the dispatcher's phone every day. They yelled, screamed, making enough noise so that we could not use our phone. We had to resort to long distance. There was about 500 miles of phone line out there, with many call boxes. I managed to cut the phone off in one direction. I found that the noise came from East of town. This left about 250 miles from which the noise could be coming. Then a tree went through the wire just East of town. The noise persisted, narrowing it down to one call box.

We had a new company cop. He not only was one heck of a nice guy, but the most gentle man I encountered while working for the railroad. I guess he had no need to be tough. He stood six five or six, weighed about two hundred eighty pounds, a former green beret that at one time taught the martial arts. I remember him winning the tri-state pistol shooting championship two or three times. Not a man to mess with.

He caught my daughter riding her horse on company property. She took a short cut along the track, instead of using the highway, and he did not know who she was. After she told me about it, I asked the cop if she could use her shortcut. I told my daughter he said it was ok for her to keep using it, but not to ride on company property anywhere else. He would not bother her. He must have scared the tar out of her, as she still had her misgivings. She said, "But, dad, he is so BIG."

The next day the cop was in town. I told him I had found the call box that the kids were abusing. He told me he would be in town for the next couple of days, to just call him on the radio when the kids came on. Some time passed, and the kid who seemed to be the ringleader of the bunch came in on the phone, doing his usual thing. The cop was only about a mile away from the call box. He told me to keep the kid talking, so he could sneak up on him. I told him I would like to scare the kid a little. The cop should key the button on his portable radio twice when he got into position.

I started talking to the kid. I had always ignored him, hoping he would get sick of it and go away. He seemed to be very pleased to have somebody to insult. Soon I heard two soft clicks come over the radio. "Kid" I told him, "I have a giant on your trail. I told him to slap the shit out of you, boil your nuts in hot oil, cut you up, and feed you to the pigs."

"You're shitting me," was his response.

"Just turn around and look." I told him.

"Oh, my God," I heard him scream. This was the end of that problem.

The Bear

The roundhouse foreman at Hog Hollow is the strongest man I have ever been around. He did not look the part. Sixty years of age, six feet tall, two hundred pounds, he would not stand out in a crowd. His shoulders were broader and his hands a little bigger than average, were about the only clues. A new drawbar for an engine weighed over four hundred pounds. He could pick one up and slide it in place with a grunt and little apparent effort.

My oldest son worked for the railroad during the summer

months. A football player, he made honorable mention on the state all star team at both guard and defensive end. He took great pride in his strength, and was by no means any ninety-seven-pound weakling. I had made a comment on how strong the roundhouse foreman was. "That old man" he asked, not at all sure he believed me.

One day the switch engine derailed. Seeing my son at the roundhouse, I walked over. He came out of the roundhouse carrying a very heavy butterfly frog that is used to re rail an engine. With a lot of obvious effort, he loaded it on a flatbed truck. The roundhouse foreman came out, puffing on his pipe, with one of these frogs in each hand. He proceeded to sling both of them on the truck with no apparent effort. "Yup," I told my son, "That old man."

One day a small stray dog showed up at the roundhouse. The foreman fed her; she hung around, and soon had a litter of pups. The pups were given away; the dog was fixed and given her shots. She kept the mouse and rat population in check, and the foreman became very fond of her. One day an official came around and forced the foreman to get rid of the dog.

Some time later a group of officials were gathered at the roundhouse. Everyone shook hands with the foreman. The official that forced him to get rid of the dog got a handshake I am sure he will never forget. Tears ran down his cheeks. I do not know if any bones got crushed, but someone said the man was still eating left-handed a week later. The foreman had acquired the nickname of "Bear" for a reason.

An Optimistic Driver

One night I got invited to a party; an engineer had turned fifty. I had to work until midnight, and arrived shortly thereafter. The place was out in the country. A car was coming down the driveway quite fast. I stopped at the entrance and watched the car go off the road just before it got to a small bridge.

Soon there was a crowd at the scene. The car stood in the creek bed, the hood pushed up by the engine which had broken free of its mounts. The rear axle and wheels were hanging on one of the large boulders that lay in the ditch. The driver must have come to the party rather early. “Stand back, boys,” he warned, “I am going to drive that thing out of there.”

Help Arrives

The bean counters came out with what was called a train performance report they wanted done for each train. They did not get done. They got nasty about it. I told the guys to see if they could get one done once in a while, to make it look like we were at least trying. The bean counters got nastier.

I wrote to them, telling them I could not expect my men to do anything I myself could not do. They wrote back, saying that these reports were to be done in preference to other duties. I wrote back telling them we would be very glad to comply, if they would furnish us with a list of trains that they wanted to see sit while this work was being done.

A time study was done. Upper management put on a full time chief and an extra set of dispatchers, splitting the territory in two. We all had to work some overtime. An office had closed and some of the men moved to our location. The jobs were filled.

The New Chief

The new chief was an excellent operating man, a swell guy away from the office. He had one glaring fault. He did not stand behind his men the way he should have. Sometimes he displayed an almost abnormal fear of upper management. Some of the men were new; some were new to the territory. Mistakes and second best decisions were made. Very normal for the circumstance, but upper management got rather nasty about it and the chief did little to protect them. It was no longer a very happy place to work.

We had a new dispatcher who was a very nice looking, clean cut, young guy. He was becoming an excellent dispatcher. We had a young gal working an operator's job. It was very evident he did not like her. He would stay as far away from her as much as he could. He was happily married, maybe she had made a pass at him, I don't know. We were making a transfer one morning and she walks in, black eye, slit lip, and her face bruised up. He asks her, "What happened to you?"

"My boyfriend hit me," she replies.

"Why," he asks.

"Because I hit him" she says.

"Why did you hit him," he asks.

"Because he called me a two-bit whore," she replied angrily

"What did you hit him with, a bag of quarters?" he asks. That ended the conversation.

Shortly thereafter a job up North opened up. I had a sneaking suspicion our office was not long for this world. Upper management would consolidate it with the Chicago office to save labor costs. I did not want to go to Chicago. For this, and other personal reasons, I decided to take the job up North and make a new life for myself. I had spent enough time in Hogback Hollow.

CHAPTER NINE THE NORTH COUNTRY

A Little Union Work

The chief rose to meet me. He was an older man with a big smile and a firm handshake, dressed in a nice suit and tie. I knew him by reputation only, and had heard he was a mother hen who took good care of his chicks. I did not know what a railroad genius he was. This fact became ever more evident as time wore on.

The superintendent was also the terminal manager. He was a fun-loving Irishman, with a quick smile. He proved to be an excellent railroad man. I thought to myself that it would be hard not to like either one of these men.

This was the fifth new territory I had to learn. It was an easy one. The office where I had first started dispatching had closed. What remained of that territory was now dispatched here, in addition to about two hundred miles of main line.

I was ready to go to work in a day or two. Although the jobs could get very busy at times, they were not overloaded. Life was good. I now held the general chairman's job in the train dispatchers union, and spent some of my spare hours on this.

One day I get a telephone call from the office chairman in Hogback Hollow. One of the men had been pulled out of service for drinking on duty. We were able to get the assistant vice president of labor relations on the telephone with us. Someone once stated that any dispatcher drinking on duty should have his bag slit, and his left leg inserted through the slit. I did not necessarily agree with this form of treatment, but agreed with the principal. I made the statement I had a hard time believing the man was guilty of this. I had worked with him several years. I knew he liked a little schnapps now and then, but had never seen him drink on duty.

The office chairman said the man had changed, and this

was no longer true. We hemmed and hawed around for a while, coming to the agreement that as soon as the chief of the newly established department of social services was convinced this man was fit to be returned to duty, he would be. I felt this was the best that I could do for the man; they had him dead to rights. I did not want him sitting in a dispatcher's chair while under the influence, in any case.

About two weeks later the phone rang. They had reinstated the man. It should not have surprised me, anybody slick enough to keep a conductor celibate for months would not have much trouble handling someone from social services. The silver tongued old rascal had concocted the story that he had a few drinks on an empty stomach while watching the World Series, forgetting that he had to go to work at four that afternoon. His wife reminded him. He hated to lay off on such short notice, and he had no trouble with alcohol.

The truth of the matter was he kept a bottle in the mailbox outside the office door and would sneak out regularly to get a nip. The story sounded strange to me when I heard it. He was not a baseball fan. Evidently the man from social services wasn't either. The day in question was a travel day. There had been no ball game.

The man was only three or four months away from retirement and managed to keep his nose clean until then. He passed away a few days after he retired. Nobody besides his doctors knew he had a fatal heart condition, not even his family. Not that it excuses it, but I guess nobody needed a diagram to explain his drinking.

Then the railroad filed for bankruptcy. The carrier wanted an immediate 10 percent wage reduction. There were bankruptcy hearings in Chicago on a regular basis. Besides attending these hearings, I often had to go in on union business.

I flew back and forth because it was cheaper than the time off, hotel, and meals involved would be if I took the train in.

At one point I think half the stewardess on the airline I used knew me by my first name. One made the comment, when she woke me up at the gate, that I was the only man she ever knew who could get an hour nap in on a fifty minute flight. I had to get up very early. I needed a nap on the way back.

I fought the wage reduction. I knew that the railroad, when formed, had been established on a land grant basis. The last I knew, still owned about two hundred thousand acres of timber out west. The land was not listed as an asset. The trustee threw some very sharp inquiring looks when he heard this. Investigation revealed the property was part of the newly established land company, a wholly owned subsidiary of the railroad, and had been overlooked.

I testified that my dispatchers could get jobs anywhere, if they could not get a decent wage here. What was the use of keeping the railroad running? When asked what I would do with the railroad if it ceased operation, I answered I would try to sell it. If unable, being an old farm boy, I would probably plow it up and plant potatoes.

We ended up making a deferred wage agreement. Anytime the railroad made so much a share, or was sold for a certain price, we would get our money. Inexplicitly, the railroad started making money. We were paid.

I normally do not have much nice to say about politicians. The trustee was a former governor of Illinois. I felt he was very fair. He handled his chore about as well as any man could. I have nothing but respect for him. We ran into each other at an airport a couple of years later. We had a drink or two together, and he remarked, "Farm Boy, you got one heck of a nice price for those potatoes." I liked the guy.

Word came out that our railroad had been purchased by another. Nothing much changed for a while, not even the color of our paychecks. I was now working the chief's job on week ends and when he was absent. I spent one day in a neighboring train dispatcher's office and made out reports two days a week. I was the second oldest man in the office. The oldest was working first shift. He had a quick temper and an acid tongue. He wanted nothing to do with chief's work. He thought he could get himself into enough trouble where he was.

A Smelly Proposition

One Friday afternoon a farmer pulling a honey wagon full of liquid manure tried to beat a local over a private crossing. The tractor and farmer made it; the honey wagon did not. It was a warm day, and the windows on the engine were open. The honey wagon all but exploded when it was struck, drenching the engine in liquid manure.

I called the roundhouse, asking them to please call the men's wives to bring some clothes over for them; I did not want them to have to go home nude. I suggested it might be a good idea for the speed tape on the engine to get lost. If someone complained, tell them it stunk too bad to keep. The local did not work on Saturday and Sunday. This was derby day; the crew in a hurry to get home was probably going like the mill tails of hell. The roundhouse foreman laughed and assured me he would see what he could do.

The conductor on the job was a rather small man of French decent. He was a deeply religious fellow who lived, not preached, his religion. In my opinion, one could not have asked for a better employee. I saw him the following week and asked him, "I have known you for several years, and as far as I know you have never lied to me. I would like to ask you a question, last Friday when you hit the honey wagon; did you say any naughty words?"

He looked at me, smiled and replied, "Joe, I am ashamed to admit it, but I said several of them." The next time I talked to the roundhouse foreman he told me he had wished he could have thrown the whole engine away, not just the speed tape.

Night Chief Duty

Things were too good to last. One day we were informed we would be getting about a hundred and fifty miles of very busy main line territory. The chief fought this move every way he could, telling upper management it would not work. The positions would be overloaded. Maintenance of way forces and switch engines would stand, waiting for the dispatcher. It would cost more money than it could possibly save.

The bean counters had an insatiable desire to fix what was not broke, and there was no arguing with them. They did consent to establishing a night chief position, if the chief wanted it. More to see if they could shut him up than anything else. He asked me if I would take the job if he established it. There was no way I was going to say no to this guy. I did not like the hours or chief's work, but if that is what he wanted, I could live with it.

I was there five days a week during the hours when the work load was at its peak. At least I had week ends off. I do not know how much good I did for the man in the dispatcher's chair. I helped answering the phone that always seemed to be ringing I did a few reports, slapped a band aid on the operation now and then, and left a nice report for the chief every morning. He appreciated having the facts in front of him, and not having to dig for the information. It probably helped him as much as anybody, but that is not the reason he put the job on.

One morning about one o'clock a young conductor came into the office, laughing about something. He picked up his orders and started to laugh again. I asked him what he found so funny. He told me about his adventurous ride to work. The engineer on the job, the oldest on the district, was a real card. The engineer made more than a fair buck. His wife was a full professor at the local university; they were both about ready to retire. I imagine he could have driven any luxury auto he chose. Instead, he always came to work in a black 1949 ford pick up he had restored. It was his pride and joy.

The brakeman on the job had far more seniority than the conductor, but would rather work as a brakeman. He had inherited the proceeds of a very successful business his father and four brothers owned. He was the only heir, and from what I heard, had added more than considerable wealth to it. He was a huge man, his fingers looked like bananas.

The engineer always picked the two of them up on his way to work. This day the engineer told the brakeman that he was sick of driving, he had to drive 200 miles more when he got to work, so he was going to have to drive. The man did, but promptly snapped off the gear shift lever on the pick up. He was using a vice grip wrench as a handle and not doing a very good job of shifting, when a highway patrolman spotted them. He stopped them because he thought they had been drinking.

"Whose vehicle is this?" asks the patrolman.

"I stole it, so I guess it's mine," replied the engineer.

"What do you mean, stole it?" asks the patrolman.

"Well, I bought it so cheap I tell people I stole it," he replied.

"Do you have insurance on it?" asks the patrolman.

"Sometimes" was the reply.

"What do you mean, sometimes?"

"Well, I don't drive it in the winter and let the coverage lapse, so sometimes I do, and sometimes I don't," he replied.

Apparently having enough of talking to this guy, he asks the driver, "Do you have insurance?"

The driver replies, "No, no I don't. I don't buy insurance; I let that to the poor people. I buy insurance companies."

Now sorry he asked he turns to the engineer "This thing sounds a little loud, what kind of shape is the muffler in?"

"Brand new," the engineer replied, "I just made it last

week. I used the best Folgers coffee can available."

Now the patrolman had enough. "Out of that truck, all three of you. Now you," he said, pointing to the conductor, "Come here."

He wanted to talk to the only apparently sane one of the bunch. After being taken aside, the conductor told him who they were, where they were going, and that yes, if they had the insurance company stock the man owned they probably could both afford to retire and move to Florida.

The patrolman now wanted to see the brakeman's driver's license. He pulled out a huge wallet, held together by very heavy rubber bands. The contents revealed not only a valid driver's license, but more cash than the patrolman had probably seen in one pile in his life. He believed they were who they said they were.

It was fish fly season, and they had been standing near a light. "Just a minute," called the brakeman. "You have fish flies on your back; let me get them off before they get your pretty uniform all dirty." Pick them off he did, he squeezed the juice of each one on the patrolman's uniform as he picked. He had a bone to pick with the highway patrol. Some months ago he had been stopped at a random check road block and forced to post a $25, 000.00 bond in lieu of car insurance. He was not mad at having to post the bond. What made him mad was he did not have quite enough cash on him and had to write a check.

One day the chief was on vacation and I was working his job. The regular day yardmaster was also on vacation. His replacement was a small man not much over five feet tall, with a rather big mouth. The first trick dispatcher had no more than started work when the yardmaster asked permission to leave about a hundred cars of grain on one of the two main tracks. The dispatcher gave him permission, but told him he was, under all circumstances, to keep the other main clear because

Amtrak was late and he needed a track to run him on.

Up until this time, each railroad used its own radio frequency, but this was the big day. A single frequency was now to be used. A big deal was being made about truly becoming one railroad. I don't know what kind of ceremony they had, or who all was there. They were making a big deal out of throwing the switch.

About this time the dispatcher noticed an approach light on his only clear main. He tried calling the yardmaster on the telephone. The line was busy. He called on the radio, the yardmaster answering him immediately. "What do I see on the Eastbound main?" he asked, "Amtrak is getting close"

"Thirty cars of coal, going up town," He replied, forgetting about the grain on the other track, he got smart. "What is the matter, don't you know how to bend the rail?" Meaning don't you know how to wrong track the train. At this moment, the switch at company headquarters was thrown.

The first words uttered on this frequency were immortal. "You four-foot, mother fucking, son of a bitch," boomed over the airways. The coal was promptly backed into the yard. I do not think Amtrak was more than slowed down. My phone started ringing almost immediately.

"I want that dispatcher taken out of service at once," the voice told me. He did not identify himself. I didn't want to know who he was.

"Well, sir, I am only the relief chief here. I am not sure I have anyone available to relieve him. Maybe we had better take this up with the superintendent."

To my pleasant surprise, he did not argue with me. I had half-expected him to say I should get my own ass in the chair, but maybe he was too mad to think of that.

I made a bee line for the superintendent's office to warn him. I was too late. He was just picking up the phone when I got there. To my relief, he had evidently overheard the conversation. He was not on a speaker phone, so I only heard half of the conversation.

"Well," he said, "I think that is a rather appropriate name for a man who runs thirty cars of coal out in front of a passenger train." They discussed the matter, a little Irish charm was used, and the man at long last settled for an apology. No ordinary one. It had to be delivered that day and was to be addressed to everyone. It started with me, then the Superintendent, up to, but not including, the Virgin Mary.

The apology they got would have been approved of very quickly by General George Patton; it was the type he gave. I had to accept it, since I wrote it. The first trick dispatcher signed it without reading it. I could tell how sincere and repentant he was. I gave him a copy, just in case he had to further defend himself. Nothing more was ever said about the incident.

All Power Is Not Created Equal

We got some new engines. Nice, huge ones, six traction motors instead of four, low gear ratios, ideal for lugging five thousand tons of coal up a hill. One small problem. The king of bean counters decreed they were to be used on our sprint trains, hauling maybe two thousand tons of mostly piggyback business. These were our crack trains, so they were to get the new power.

This made about as much sense as buying an automobile with a huge eight cylinder engine and running it in low gear, wide open, down a freeway. The roar of the engine was deafening; ear plugs were in order. The engines sucked huge amounts of fuel, and could barely make track speed.

All of the power generated with no place to go, caused the engines to vibrate violently. The radiators, which should have lasted many years, were ruined in about five months. The pipes leading to and from them shook so badly that it caused metal fatigue, they leaked like sieves. I understand that General Motors, who had manufactured the engines, told them they were idiots when they got complaints.

It did not take long before we again had two smaller engines on the sprint trains. Both engines were used only on the hills, the lead unit taken off the line in flat territory. Now it was nice and quiet in the lead unit, and top speed was more than ample. I do not know what further damage had been done to the new engines, but it had to be hard on them. At well over a million dollars each, it had to be an expensive mistake.

After the acquisition, a transfer track was no longer used and a switch crew delivered cars directly from one yard to the other. There was an engineer working for the acquiring road who had spent his entire career working in a flat yard. He maybe had never run an engine faster than ten or fifteen miles an hour in his life. The trainmaster rode with him on his first trip to the other yard.

The man got to the point where the track started down a long hill. He took one look at the hill and set the air, hard, stopping almost at once. He told the trainmaster he had just retired. The trainmaster took the cut of cars down the hill. When they got to the East end of the yard they met a sprint train. This train was on the adjacent track, going about sixty-five miles an hour. The now retired engineer refused to ride the engine back, walking to the yard office and calling a cab to take him instead. He had seen enough.

A Titled Man

We had a string of young college graduates pass through the office. They were given the title of assistant trainmaster. They were there to observe and learn. Then they would be placed in one of the crafts, mostly to the clerks, to get a little experience before they would be sent out as trainmasters. They were, without exception, nice, sharp, young people. They stayed about two weeks and left, where to I have no idea.

One day a guy showed up around the office. He looked a little older than most of them. I did not pay much of any attention to him. I heard his name mentioned on the company radio, I had no idea that he was anything more or less than another assistant trainmaster.

One evening the trick man was away from his desk. I heard a radio conversation. The head end of one of our trains was on his way up town, pulling out of the yard, through the entire plant, to get where he was going. The train had trouble and now sat blocking both our main tracks, both of the BN main tracks, our lead, and the entrance to the Duluth Line.

I had heard this fellow out in the yard on a mobile unit, talking to the switchmen, and wanting to get our train moving as soon as possible, I called him on the radio, addressing him as an assistant trainmaster. I politely asked him to please help the crew on this train.

The answer to my request was a tirade of screaming. "I am not an assistant trainmaster, I am the assistant terminal manager," and it went on from there. The BN dispatcher had several trains that would be standing soon if the problem were not solved. He was not one to put up with much nonsense. He broke in on the radio and informed the guy he did not care what his title was, just to get his lily white ass over there and do what was asked of him. The problem was soon solved; it took them only about ten minutes to replace a broken air hose.

When the train got moving, he called the BN dispatcher, and told him, "I am the assistant terminal manager, not an assistant trainmaster."

The BN dispatcher replied, "I am sorry to hear that you have been demoted, sir." That one really pissed him off. For some reason the man blamed me for starting this whole smelly ball of wax rolling. He hated me ever after that. I left a little note for the chief, just in case.

The next morning he came in to see the chief, his boss in tow. I was not there, so I can only tell the story the way it was told to me. The man complained to the chief about the lack of respect he got from the train dispatchers, demanding the chief do something about it. The chief told him with a big smile on his face, he was glad he had stopped in, that he wanted to talk to him about this very issue.

"How many of the dispatchers did you bother to introduce yourself to?" he asked. The man admitted he had not done any of this. "How can you expect respect from men that do not even know who you are?" he asked "And further more, did you bother to introduce yourself to me?" he asked. "I have never experienced such rude behavior from an official in the forty or so years I have worked for the railroad. You should be ashamed of yourself." The chief went on to a laundry list of things the man should have done, not done, or could have done better. All the while the man's boss is sitting behind him with a smile on his face. He had seen the chief in action before.

To make a long story short, the man not only got a blistering lecture from his boss, but had to go to each dispatcher, introduce himself, and apologize for not doing so sooner. Even though I told him I had no problem with this and was sorry I had not used his correct title, there was still hate in his eyes. I think he also blamed me for the blistering he got and for having to apologize. His boss had lost one darn good trainmaster in exchange for what he got. He was not a happy camper. His boss would look for excuses to chew on him, and he would find plenty.

A few days later, the centralized traffic control system crashed. The assistant realized trains were standing, and called each of them, switch jobs, locals whatever, to ask if they were moving. Nobody was. He called the BN dispatcher who ran the system asking why none of the trains were moving. He used a harsh voice as he was still mad at him. It was the same man who was working previously. "Because some ass hole in a mobile unit has had the airwaves tied up" was his answer.

The assistant called the BN chief dispatcher, insisting that something be done with this man. The BN chief not only chewed on the assistant for bothering him and his dispatcher and monopolizing the airwaves, he wrote a note to the Superintendent complaining about this very issue. I imagine that he got another blistering. The BN office evidently had a good mother hen, too.

Sometime later, we got a very heavy rain. The assistant was not in his office. The trick man and I managed to get people out to patrol the track, including the terminal. There was a washout at the West end, not long, but about eight feet deep. Amtrak was due in about three hours and had to use this track.

I called the superintendent at home to advise him. He had stepped out to buy some cigarettes; his wife informed me. I told her about the washout and to please have her husband call in.

I heard the door slam; the assistant was back. I asked him if he had heard about the wash out, and if we had any rock in the yard. He replied, "I am not the fucking yardmaster."

I went back to my office and called the yard. The yardmaster, an old head, was way ahead of me. "Yup, four cars on the top end of the rip, I heard about the washout and sent the switch engine to get them."

I thanked him and heard the phone ringing in the assistant's office. I had a very good idea of who it was and what he wanted. I stood outside the door for a moment to let the boss scream at him some more for not knowing if we had any rock on hand or not. Then I walked into the office and said, in a loud voice, "If that is the boss, please tell him there are four cars of rock here, and the switch engine is on the way after them." Now I could hear the superintendent screaming at him for making the train dispatchers do his work for him. He was not on a speaker phone, but he was talking so loud I could hear him anyway. As a result, for trying to do him a favor, he hated me even worse.

We had a new dispatcher working on a neighboring subdivision. Our trains ran on both subdivisions. When an inbound train got into town the assistant would grab the delay report and call the new dispatcher, a gal, to chew on her about something. I'm not saying this gal was any great train dispatcher, with about two weeks experience she couldn't be, But, I thought she was doing very well, all things considered. I asked the man to leave her alone; she was doing more than ok. A little encouragement and a kind word were maybe what she needed.

If you leave out the profanity, the assistant did not say much. I called the gal. I told her she was doing ok, to ignore him and throw the phone in the wastebasket when he called. She evidently could not do this. He kept chewing on her, and she quit.

I wrote a nice, polite letter to the superintendent of the neighboring division with a copy to the chief dispatcher there. I said I felt it was a shame that after the railroad had spent a lot of money training her, we had to lose her. I stated that, as both the general chairman of the train dispatchers union and as an assistant chief, I thought it would behoove us to see if she could be talked into coming back. I thought she might become an excellent dispatcher if given the chance.

I also stated that from what I had seen of the assistant's behavior toward me and other people, I could see where she might want to quit. I laid this part of it on rather strong. I asked them to please see what they could do. I did not know the superintendent there at all, but I had known the chief since I was in high school. He was the station agent in my hometown.

He could read between the lines, and knew exactly what I wanted. It was like saying 'sic em' to a vicious dog. If I had written "Crucify the SOB" across the chief's copy, I could not have made myself any plainer. If the assistant wanted to hate me, I would help him a little.

The neighboring chief called our superintendent, complaining bitterly that his assistant had chased away his best prospective dispatcher. He had to pay overtime. He suggested he call the assistant's mother to come down, take him into the basement, and wash his mouth out with soap.

The neighboring superintendent wrote the assistant a nasty letter, with a copy to his boss. The assistant now had to call the new dispatcher, apologize, and ask her to please come back to work .I was told she refused to talk to him, and never did, to my knowledge, rejoin the train dispatchers ranks. This further infuriated both the neighboring chief and superintendent. They did not make life pleasant for the assistant.

It was Thanksgiving; the superintendent was on vacation. The assistant had taken his place. If he had been smart, he would have sat back, not troubling trouble until it troubled him. He would have let the chief and trainmasters run things, and had a very nice week indeed.

Instead, the assistant was working eighteen hours a day, accomplishing little, except getting more people angry with him. We ran four road switch jobs out of the yard that serviced industries on little branch lines close to town. Instead of just not running the trains on the holiday, he decided to make himself look good by not running anything until the following Monday. He sent Chicago a message about how much money he would save.

Things did not start biting him in the posterior until Sunday night. There were around two hundred and fifty cars in the yard that should have been gone. Room was getting scarce. Switch engines were working overtime because of the congestion.

Monday morning he compounds the felony by grossly overloading all the locals in an effort to get the cars out of the yard. The little switch engines did not have enough power to make the hills. They stalled and having no place big enough to accommodate half of the train, had to put it away a few cars here, and a few cars there. Now everything had to be done long division. There was not enough room anywhere to do anything quickly. Aware of the fact they would be out there more than twelve hours, the crews were in no hurry to go anyplace.

The chief had been trying to talk the assistant out of doing this, but to no avail. Sometime later Monday morning the chief went into the Superintendent's office. He had a big smile on his face and some paperwork in his hand. "Look at what we have been blessed with," he told the man.

The superintendent looked at the facts and hit the ceiling. "We will be out there twenty-four hours a day, for damn near the whole fucking week," he proclaimed.

"I know," said the chief, with a smile. The assistant was now holding the top spot on the chief's shoo-shoo list. A very uncomfortable place to be. The chief was a master at giving you something you should not have, and then making you sorry, very, very sorry you had asked for it. I heard when the debacle was complete, the chief drew up an estimate of his own on how much the "economies" had cost, sending a copy to Chicago and one to the Superintendent.

I do not recall the exact figures, but the estimated savings were three or four thousand dollars. The costs were up toward thirty thousand.

The assistant came charging into the chief's office demanding why he had done this. The chief smiled at him sweetly and told him "I thought you would appreciate having the figures, just in case you decided to do the same thing again next year."

There is some sort of an oriental torture known as "the death of a thousand cuts." This I am sure the chief did not favor, but used something more like "the life of many needles" to accomplish his goals. I think the basic difference is the guy with the cuts, at least at some point in time, is allowed to die. The guy getting the needle lives indefinitely, but no more comfortably.

And needling was this chief's specialty, as the assistant was soon to discover. If needling were an Olympic sport, the chief would have had so many gold medals hanging on his wall that the wall would have collapsed.

The chief did it in such a nice way. Gosh, gee, and darn were strong language for him. To illustrate the point, I once asked a conductor how things were going for him. "Fine," he replied, "Except the chief had me in his office here the other day for a pep talk, as he called it. He smiled the whole time, never used one four-letter word, and I didn't think too much about it, until I got to the end of the hallway. Then, I felt like my ass hole was missing, I looked back to see if it had dropped down to the floor, but it wasn't there. But the really strange part of it all was that I ended up being mad at myself and not him."

I am sure that the assistant experienced this feeling many times in the next while. The chief also had his little ways of getting things accomplished. We would bootleg cars of grain around on coal trains, time freights, or whatever. The computer now kept track of cars and we had to come up with something to keep the computer happy. It had to know where the cars were, when moved and on what train.

There was a burned up engine sitting near the roundhouse. It had been in a gasoline fire a few years ago. There was not much of it left. The computer recognized it as a valid engine number, so we would run extra trains, on paper, with this engine to keep the computer happy. I bet we made more miles with it after it had been destroyed by fire than it had it had made before.

The superintendent knew what we were doing. He got a big kick out of us using foreign road engines and fuel to haul our business around. He was not about to rat on us.

Winter came, and with it, snow in Duluth. Lots of snow, almost three feet in about as many days. It quit snowing and blowing one Friday evening. Knowing that we always had a lot of cars to move after every storm, I called the yardmaster in Duluth. He was an old head, had lost a lung in the Battle of the Bulge, and was not prone to take too much guff from anybody. Evidently the assistant had been chewing on him for too much overtime on his switch engines, because he hated him with a passion.

I asked him what he had in sight to come South. "Let's see," he said. "The regular trains, two one-hundred-fifty car trains of solid lumber on the way down from Thunder Bay. The transfer tracks have not been pulled in three or four days, maybe about a thousand cars." I thanked him and hung up.

I arranged for two extra sets of engines to move North on the next train, clearing the move with the control center. We ran two less East trains on Saturday night than normal. The power would be back by Sunday when we would need it.

There was a nice new kid working at the control center. He told me that by all means, if I thought we needed the power up there to send it. Later the assistant sees all these engines lined up to go North and calls the yardmaster in Duluth. When asked, the yardmaster replies. "I have eight cars in the yard." This statement was true.

The assistant called the control center, telling him the engines were not necessary. He has the crew take the engines back to the roundhouse, and calls me on the radio. He screams at me, telling me I was insane for sending them, and asking me why I did. My reply was simple. "Because we need them," I told him. He snorted at me, and that was the end of the conversation. I left a note for the chief and the superintendent, telling them of what the assistant had done, and sorry, I was through arguing with him. After the two regular trains departed Duluth that day, there were still over nine hundred cars on hand.

The superintendent developed a new routine that lasted for about the next ten days. He would come to work, pour himself a cup of coffee, light up a lucky, look at the Duluth on hand report, and go looking for his assistant with blood in his eye to give him another ass eating. This suited me fine.

Then something else happened. The young fellow on the control center was evidently chastised for letting the assistant take the extra power off of the train. Now he would not make a move with an engine without consulting me. Every time the assistant so much as wanted to breathe on an engine he was told, "I'll have to check with Joe, first."

Now the assistant really hated me. I heard when the report showing all the overtime the Duluth switch engines worked because of the congestion came out, he again had his ass royally chewed. He had further reason to hate me. The old yardmaster had done a masterful job of hanging him out to dry.

The assistant's name was not only on top of the chief's shoo-shoo list. It had a couple of gold stars by it. One night, the assistant had the night off, but was on call in case of an emergency. A man got hurt, I tried to call him, several times, but there was no answer. Thinking he was out awful late, I had a trainmaster handle the situation. There was no problem. I left him a note, telling him what happened, and that I had tried to call him. I went home from work, to bed, taking the phone off the hook as I often did.

The assistant shows up the next day. He finds the note and gets on the chief because he could not get me, all bent out of shape about it. The chief informs him that I am required to get my sleep, so forget about it. He starts screaming that I should have tried to call his cell phone number. "The only reason I can think of that you would not answer your home number but would answer your cell phone number is you were not sleeping in your own bed," the chief tells him. It turned out that the assistant's telephone had malfunctioned. The chief asked him, many times, sometimes over the radio, "Shall we call your cell phone, or will you be sleeping in your own bed for a change?"

More Union Work

It was about this time I made my last appearance at a standard investigation. We had a foreign line local tied up on our property. The train went one way one day and back the next. The conductor on the job was a real worker and had acquired the nickname of "Go," and go he did, sometimes too quickly. The train had not been on the road for more than five miles when he had to meet an opposing train.

The siding there was seldom used because it was too short for the average train. Instead of having what is called a power switch so equipped that it is thrown by the train dispatcher, it had a hand throw switch. Once the switch was lined and locked for either the siding or the main track, the dispatcher was able to give the crew the signal to proceed.

Seeing a yellow approach signal when they were getting close to this siding, the crew suspected they had to meet a train there. They managed to contact the opposing train on the radio, and he was getting rather close. Getting a red signal and a yellow S light meant they were to take the siding. They quickly lined the switch into the siding. Instead of waiting for the signal to turn the right color, as it would have when the dispatcher got to it, the conductor gave a hand signal to the engineer to proceed into the siding. This was safe enough, they could see to the other end of the siding and the switch could not be moved by the dispatcher.

They pulled into the siding and lined the switch back. By the time they got to the other end of the siding, the opposing train was showing up. When he went by, they lined the switch for the main; they got a green signal and left, lining the switch back for the main track behind them.

The dispatcher, despite his now overloaded position, managed to notice them going into the siding without a signal. He was not too happy with them. They had some cars to spot a station or two down the track and needed to get off the main through a power switch with a signal on it. The dispatcher threw the switch for them, and bang, they went through the switch before the signal had a chance to turn color. Now angry with the crew, the dispatcher hollered for the chief to come here.

I think if it had not happened back to back, the dispatcher would have been content to do a little chewing and let it slide. Left to his own devices, I think the chief would have arranged to give the conductor one of the pep talks he sometimes gave, and let it go at that. This, however, was not to be the case.

The assistant, using the copy machine near the door, heard the dispatcher telling the chief what had transpired. In his usual own loving way, he immediately formally charged the conductor for the two perceived rules violations, and the dispatcher for not immediately reporting the first incident.

I was no longer the general chairman, but was asked to represent the train dispatcher. I, in no case, would have declined to do so. The source of the charges made doing so all the more appealing. In no way, shape, or fashion was I going to allow the person pressing the charges to conduct the investigation. I would scream extreme prejudice and could prove it.

I guess labor relations knew me fairly well by now. Before I had even protested, I got a letter from them saying they had disqualified the man. Of course, they wanted something in return. It seemed the local superintendent was too busy. Since it formally required at least someone holding the rank assistant superintendent to hold an investigation on a train dispatcher, would I waive this requirement, and allow two of the local trainmasters to conduct it.

I wrote back to them. In an effort to save the railroad the expense of importing an assistant superintendent, without setting any precedents, that, just this once, I would waive the requirement, and then only because of the quality of the two men involved. I went on to say what fine officials I thought they were. How intelligent and knowledgeable they were. I polished their halos as best I could. Hell, I liked these guys; it is not that often you can make your friends look good.

I knew there was a rule in the book someplace allowing a train to clear the main track at an outlaying point when communication with the train dispatcher had failed. The chief knew it was in the book, the dispatcher was rather doubtful. We looked. And we looked and we looked. All to no avail.

A traveling engineer, working on a case involving protection of the rear of a train, found it for us by accident. For some reason the rule was buried in the middle of several rules regarding protection of the rear end of a train. I still owe him another thank you.

We got to the investigation several minutes early. The trick man, me for the defense, and the chief as a witness. We sat down. The first thing the trainmaster said to us was "Well, I see they sent me a bunch of rookies to deal with again. Between the three of you, you have barely one hundred and twenty years of railroad experience." That drew a laugh.

We were handed a "History of Incident" report showing the facts, as known, surrounding the incident. It was very well drawn up, complete and accurate in every detail except one. The presiding officer stated he had talked to the conductor on the radio at the office where he came on duty on the morning in question. There was no radio there.

He knew I knew there was no radio there. I knew he knew there was none. He had just presented me with a hole a drunk could drive a semi through. My driver was sober and driving a Volkswagen

The investigation was called to order. "Does anybody have any problem with the history of the incident report?" was the trainmaster's first question.

I raised my hand. He nodded at me. "I find it extremely well prepared and correct in every respect, except that I think you may have talked to the conductor on the telephone and not the radio. It sometimes happens, but it is a fluke, if we have radio reception in this area. Are you sure you did not use the phone?" I asked.

The trainmaster replied, "Oh, I used the phone, I must have forgot. Yes, I know there is no radio reception in this area."

Forgot, my sore ass, I thought to myself. My Volkswagen was long through the hole, disappearing in the distance.

"Does anybody have anything else before we began," the trainmaster asked.

Again I raised my hand. "In an effort to save some time, I would like to ask the conductor a few questions before we proceed, if nobody has any objection."

"I have none," the trainmaster replied. The union representative for the conductor was somewhat hesitant, but gave his approval. The conductor took the stand.

"I see you have a copy of the current consolidated book of rules in your hand. Would you please open it to page one hundred thirty seven, rule ninety nine, third paragraph, and read it out loud, please." I said.

He did so. Now I was getting a big smile. "Would you say you complied with this rule in its entirety when you cleared the main track on the day in question?" I asked.

I had no doubt what his answer would be. He had just

exonerated himself of half the charges. Ignorance of your innocence shall not be construed as guilt, I thought to myself. I got a nice smile from his union representative, too.

This subject was never revisited. There was no need to. Investigations are won on the knowledge of, or the ability to interpret a rule in your favor. They are not won by silver-tongued oratory. Which is a good thing, because I flunked oratory a long time ago. If a jury had to hear me talk for very long, they might hang me.

The conductor later testified that the sun must have hit the signal at the other location, making him think it was green. He was barred from our railroad for a week or two as punishment. He went on a job nicknamed the "dirty underwear." It got its name from the time during World War One that this job went to a near by military installation, hauling dirty clothing to a laundry down town and clean clothes back. Nicknames die hard. His hours were not quite as good, but the job got some overtime, so he got an involuntary raise.

The investigation cost the railroad less than a thousand dollars to hold. It was a cheap one. The chief could have handled it with a little pep talk in about five minutes. The results may have been better. It would have saved a lot of hassle for several people. The only good part of it was that it made the assistant look somewhat stupid when the man was found innocent.

Before the investigation closed, I was allowed to make a statement. It was simple. "The first incident did not need to be reported, there was no rule violation. The second incident was reported the moment it happened. The dispatcher is guilty of nothing."

The standard procedure when closing an investigation was for the presiding officer to ask "Have you received fair treatment, and was the investigation conducted in a courteous manner?"

The standard answer was "Only the national office can

determine that."

When asked I replied, "Extremely so, it is hereby noted, and very, very, much appreciated."

I was not about to feed my friend a bunch of bullshit, when the truth would do even better. We both knew what the outcome would be. He was my best witness. I just tried to polish his halo a little more. He deserved it. He was one of the best. This is the only investigation I have ever heard of where the presiding officer was prejudiced toward the defense. I don't think that the trainmasters liked the assistant either. Not only that, I didn't have to work very hard that day. The chief had an even easier day. He was never required to say a word. He did smile now and then.

I had to appear at an investigation in my own defense only once. There had been an ice storm. The centralized traffic control system was not functioning. I gave a crew permission to line the switch manually and proceed by the signal at restricted speed. It was raining and the engineer, in an effort to keep his head break man from getting wet, got too close to the switch. The slack ran in, shoving the lead truck of the engine into the switch just enough to bend the switch rods.

The crew reported the bent rods to me. I got a section man who lived near the site to go over there and straighten the rods. The train got about a thirty or forty minute delay. The trainmaster must have been up early, he heard the conversation on the radio and called a standard investigation. He was not aware that he was not authorized to do so, but we let him go ahead.

I was asked if I told the train crew to stop. I replied I did not remember the exact conversation. I did not want to make it any harder on the train crew than I had to. I did not tell them I told the crew when they were ten or fifteen miles from the location that the CTC was out, and that they would have to stop. I did not too much care what I told them. I knew the results of the investigation would be thrown out anyway.

They gave me a record mark for not telling the crew to stop, but had to expunge it when they found out a trainmaster held the investigation. A fact the union delighted telling them about. They could not try me twice on the same charge, so all was well. The train crew got a minor slap on the wrist. I would imagine the investigation cost about two thousand dollars to hold. As near as I can figure, the section man I called out got $11.50 worth of overtime.

A pat on the engineer's shoulder, along with a "Hey, next time try to be more careful" would have worked just as well. I was going to perform that service anyway, for free, the next time I saw the engineer involved. The incident would have become just another lump under my carpet.

The chief where I held my first permanent dispatcher's job once told me, "Kid, learn the rules well, that way you will know exactly what you are violating." It was very good advice. To my way of thinking, the carrier would have been far better served if they had bought a nice, thick, juicy steak for the section man who fixed the switch very promptly, and fired the trainmaster. He is the one costing them all the money.

The carrier was very good at shooting himself in the foot, with all the practice he had, he should have been. As an example, the railroads got a subsidy for keeping Amtrak on time. We seemed to always get the award for doing the best job of this. The reason we always won was maybe we were the only railroad that tried. There is no money to be made moving people, so why delay good freight business that does make money. Left to my own devices, a passenger train was little more than a siding cleaner any time he met a freight train. However, we had instructions not to delay Amtrak. By and large, these instructions were followed.

We had an engineer who was a master at making up lost time. He was past retirement age, raised prize roses in his spare time, and everyone knew him by the name of Boot. His judgment was as good as his ability to make up time. He would go like a scald cat where the track was fit, but was known for all the messages he sent regarding bad track conditions. "Track out of line three poles West of MP 284, getting worse, going 20MPH under prescribed, still getting bad throw," is an example of what he might say. He could sense a bad track condition before any track patrolman could spot it; his engine rode much differently than a thousand-pound motor car. He missed next to nothing.

The Westbound main track beyond where he got on the train had just been renewed. The only remaining task was to shape and smooth the track shoulder. The track was as smooth as glass. There was a shoulder shaping machine working in the area, the machine itself on the Eastbound main track, the head of the machine working on the shoulder of the Westbound. He had a train order to look out for this shaper working, but when he got to this area the prescribed time had expired by about twenty minutes.

He left right on the line-up figure, about twenty-five minutes behind schedule. There was no sign of the shoulder shaper in the area that the train order prescribed. He reached a point almost a mile beyond these limits to find the shaper still working, the heads of the machine fouling the track that he was running on. I think that the speed limit was seventy-nine miles per hour; he was going something north of a hundred. I don't know where the shaper head ended up, I imagine someplace close to the county line, if not in the next county.

He struck something that should not have been there, beyond the time that it should have been on the track at all. They wanted to give him an unpaid three month vacation. He had done nothing he had not done, every trip that he made, for the last five or ten years. Everybody knew what he had been doing, but ignored it until they needed a scapegoat. He told them if they wanted to get rid of him, he could fix that. He promptly retired.

Management quickly found out he could get along without them far easier than they could without him. They not only lost his great ability and experience, but now the engineers, angry with the raw deal they thought he got, made double darn sure that none of them were caught speeding. They slowed down every chance they got, moving more like angry bulldogs than greyhounds. We still gave Amtrak every break, at the same costs, but now did not collect nearly the subsidy.

There is an old saying that one does not want a drink of water until the well runs dry. We had some very thirsty people at the subsidy well. As for the engineer, I hope that he enjoys his retirement until he is as old as how fast he was going. That his roses always bloom, and that he resides in Hog head heaven forever. He was one of the best.

The year was about 1983. I worked third trick at a neighboring dispatcher's office one day a week. The territory was short, but very busy. I had one operator who was a Radio Morse man in World War Two, been a relay operator for years and another who was also an excellent telegraph man. All three of us liked to use the telegraph wire to put out train orders, and did so every chance we got. This turned out to be the last time I used the telegraph wire.

The Radio Morse man did not use a bug, he sent by hand. I did not like hand sending, I never heard enough of it to get used to it. This man was the exception. It was like hearing Pavarotti sing, or watching Jordan play basketball. It just did not get any better than that. He had a golden fist, by far the best I ever heard.

A bunch of us would get together for coffee and B. S. after work every now and then. The other operator had been a demolition expert in World War Two, and now had a sideline of digging basements and bringing down buildings with explosives. He often had to leave early. With a big grin, he would announce that he had promised somebody a blow job that morning.

An Old Friend Reappears

A couple of Wisenheimer young dispatchers someplace told the chief rules examiner, the same fellow who wrong tracked a train around lump-lump to get him wet, he could not write a rules exam they could not pass. The next exam was a beauty. It played with your mind, first getting so complex and twisted that you could barely, if at all, figure it out. Then asking a straight forward question that looked so simple that it must have some kind of a trick to it.

An ex rules examiner now working a dispatcher's job for a few months before retiring was the only one to pass it. But just barely. I understood that he spent about seven hours chewing his pencil, and cursing the examiner and all his ancestors. I gave up on passing a little more than a third of the way through the test. I continued only because I was expected to, and wanting to see what further diabolical mind twisters he had come up with.

There was a comment box at the bottom of the test. I wrote, "This test was written by a brilliant, sadistic mind, one so evil that it would wrong track a train around a trainmaster with the explicit purpose of getting him wet. It's only viable purpose is to prove that a fool can ask more questions than a wise man can answer." He laughed like a maniac when he read it, but he flunked me anyway. This is maybe what he was laughing about.

Before long we had a new exam, sort of an easy one. We were not supposed to get paid for taking an exam the second time, but he made sure we did. He did not want to piss us off, just wanted to teach a couple of guys a lesson. I heard they both scored less than 50 percent, ninety was passing. He proved his point. That was the only rules test I ever flunked.

The Consolidation

The news came out that all dispatchers offices were to be consolidated under one roof. Spring came, they were fixing up one of the tracks east of us with the idea of tearing up the other main, leaving only portions of it to be used as long sidings. A huge gang, close to one hundred workers, was to perform this chore, using some very high priced rented track equipment.

One day the first trick dispatcher, already overloaded, was unable to get the extra gang out to work. The shit hit the fan. "Well," replied the chief, when asked why this happened, "I had to make the choice of getting the gang out there, or running trains. I chose running trains, because without the trains running, there will be no money to pay the gang with. I told you a long time ago that this would not work. People sit, every day, waiting for the dispatcher to get to them. Now that they all are sitting in the same place, it finally becomes obvious to you. And the same thing is going to happen again." It took three days to transfer dispatching of this territory to the neighboring division, the location of the proposed consolidation. Some of our dispatchers went with the territory.

The assistant came into the dispatcher's office one day and became abusive to one of the dispatchers. The chief promptly barred him from entering our office without asking permission. He told the chief he was the assistant terminal manager and this was part of the terminal and he would go where he damned pleased. With a big smile, the chief informed him he did not think so. He pointing to the threshold, and informing him that was where the terminal ended and our office began, to never, ever, come beyond that point again without permission. He took the matter to the superintendent, who informed him he got what he deserved for his bad behavior, and he was to honor the chief's request.

The chief could have just as easily barred him from the office entirely, but would rather he suffer the indignity of asking, just like a third grader asking the teacher to go potty.

In the remaining time I worked there, the assistant never asked me, even once, if he could come into the office. Instead, he would stand in the doorway, head extended, looking very much like a goose searching for a bug. One had to go through the chief's office to get to the dispatchers office, so he could not bother him, either.

The chief was going on vacation and asked me to work his job while he was gone. I don't think I would have said no if he had asked me to work for nothing. When the assistant heard I was going to work the job, he went ballistic. He did not want me to get the extra few bucks, or have the honor of working the job. He raised all kinds of hell, putting pressure on the chief.

I heard about it and told the chief to give him his way, I would get six extra days of overtime out of it. He was not hurting me. I was now working first trick, my night chief's job having been abolished. I told the chief I would try to watch the kid who was going to work the job, help him where I could, and I did not see any problem with it. The chief just smiled.

The first morning the chief came back to work, I don't think it was yet eight thirty when the needle came out. A crew should have been deadheaded to the West end the evening before to get rested for a coal train which was coming.

I had meant to remind the chief that it should be done, but for some reason it slipped my mind. Second trick and third trick did not catch it either, it really not being their job. Hitting the brakes at the doorway and sticking his nose inside, the assistant screams at the chief, "There is no crew on the West end getting their rest for the coal train."

"Gee, that's funny," says the chief, "that never happened when Joe worked the job."

The assistant ignored the first salvo, but should have quit.

"What are you going to do about it?" he asks, half snarling.

"Nothing," don't worry about it, when Joe gets a good bead on that coal train he will deadhead a crew out there and bring them back on continuous time. "Of course, the switch engine will have to go get them. That will cost a couple hours of overtime." Switch engine and overtime used in one sentence were the last things the assistant wanted to hear.

"Well, Ok," he tells the chief, "But that don't make me happy."

"Me either," replies the chief, "I am very disappointed in you."

"Huh," says the assistant.

"Why yes," said the chief with a smile. "You knew there was an inexperienced man working my job, why didn't you check to see this was done?"

"That's not my job," he replied.

"Anything that happens on the railroad is a good official's job." "I had high hopes for you at one time, thinking you might become one, but now I am starting to wonder." "And indecently," he continued, "Joe thought it was awful nice of us not to force him on my job when I was gone. He was able to pick up six days of overtime this way."

Not being that much of a masochist, the assistant retired to his office, red-faced, pissed off, and sweating to lick his wounds. He had just had a very bitter taste of the chief at his finest, giving the man exactly what he wanted, and then making him wish he had never asked for it. He did not know it yet, but he had just started to pay.

Payday came, and the assistant was greeted with the words, "Payday, I imagine Joe is laughing at you."

Those six days of overtime he got will really help his guarantee when the office is moved. The five years of guarantee money will really help Joe's pension when he retires. Joe's grandmother lived to be almost a hundred. Joe will probably be laughing at you every time he gets his pension check for a long, long time.

All of these phrases remained in the chief's needle, to be used at his discretion. I think he reminded the man of these facts every payday until he retired. The chief came into my office, worried about me. I had laughed so hard I choked.

Pin Lifter on left, Hose underneath, Knuckles in center and Drawbar is at the ends.

CHAPTER TEN THE LAST RESTING SPOT

The big day came, and the offices had been moved. My chief remained at his old post. The carrier let him remain there as a coordinator, at his old salary, until he was old enough to retire, about six months from then. They got a good deal.

In my opinion, they would have gotten a better deal if they would have doubled his salary with the option he could remain there until he was a hundred years old, if he so desired.

I bid in a first trick position on what had been a part of the acquiring roads territory. I had my sixth territory to learn. This did not take me too long. It was not an over-loaded position.

The old chief that had this territory moved with it. He had only about a year to work. I knew him from the days he had spent as a union official, and liked him. He was a very good mother hen, having a fierce loyalty to his men. He would be missed.

This chief was not quite the genius in dealing with the officials as my old chief was, but he had his own ways. We were short of dispatchers, as usual. He put up a big sign in his office, labeled "Dispatchers Extra Board."

There had been a horrible accident on the BN a short time before this. A young dispatcher, I heard, had been forced to go to work before he was ready. During his second night on duty, he ran two coal trains together. The accident killed some crew members. It also destroyed about six million dollars worth of engines, and who knows what else. The young dispatcher's name was the only one on this extra board. Management made the chief take his sign down, but they got the idea and hired some people.

A New Set Of Boys

The acquiring road had always been a profitable one. Now I could see why. Things were well maintained. If the crews elsewhere on the railroad preformed like these, I could see where they made money. I became very fond of working with them. With one small exception, they were all very capable and cooperative, true professionals.

One day, after I had been there a while, I noticed that most Eastbound trains were losing about ten or fifteen minutes between about eleven in the morning and midnight or so on almost every trip. They would do a little better than expected more than half way over the territory, fall down fifteen minutes, then make up a little time again. I could not figure out why.

A signal maintainer was in the office. I was writing something down for him. My pen went dry, and he offered his. It was an unusual one and attracted my attention. It was from a pizza parlor. He noticed me looking at it, and informed me they had the best pizza anywhere. They had just opened recently, and it was right alongside the track. The address was in the area where the trains lost time.

A light dawned. The next time the union representative for the trainman was on a train getting close to the area, I called the pizza parlor. What proved to be the owner's wife answered the phone, and I identified myself. "Do you have an order there for a train crew?" I asked.

She hesitated and asked "Is there a problem with that?"

"No, but, if you don't mind, I would like to pay for the pizza if you would be so kind as to tape a note to the box for me."

"Sure," she replied.

I gave her a credit card number and included a nice tip. The message read, "Enjoy your pizza, but please talk to me about this." I had her sign my initials to it. The train was not into the terminal very long when my phone rang. It was the conductor. "Hey," I told him, "I don't mind if you guys grab a snack out there, but I sure would like to know about it."

"Little hard to do without somebody hearing us, and we

might get our bonus cut for not eating," He replied. We came up with the idea that if they were not going to stop they should ask, in an inquiring voice, if this was a nonstop flight. I would tell them of any trains they would have to meet and know that they were not going to stop. It worked well, as most of them did stop.

This was the best few bucks I had ever spent. It told the crews that I would cover their buts for them a little, but it also gave them the idea they would not get by with too much, either. I managed to cover up a couple of very rare little goof-ups for them. They soon would do anything I asked of them, with a smile in their voice. They were a bunch of top notch guys. Even if they used cell phones to order pizza. Well, I guess if I had been on duty for several hours, I would have appreciated a snack too. They got sick of it in a few months and things went back to normal.

From time to time, I would call my old chief, just to see how he was doing. He seemed to be having the time of his life. His work load was now much lighter, and he had more time to devote to his hobbies. His favorite being needling the assistant. He delighted in telling me, in detail, of the latest fire he had built under his posterior. Practice makes perfect. He seemed to be getting a lot of it in. One day he told me the assistant had a different job. He now had far less occasion to talk to him, but still got a lick in now and then.

All employees were urged to attend a meeting. A big ball room in a down town hotel had been procured. There was quite a crowd there. The assistant, his wife, and his boss stood there. A young woman, a rather nifty looking red head, walked up to them. A big smile on her face. She threw her arms around the assistant's neck, planted a big smooch on him, and told him. "Big boy, whenever you get to town and want to do some more heavy-duty fucking, just call me." She leaves smiling and waving at him.

The assistant had just met the young dispatcher that he had taken so much delight in chewing on. I don't know what the assistant's boss thought of all this, but I heard his wife almost divorced him. I imagine that, by the time it was all over, he knew exactly how it felt to be unjustifiably screamed at. Too bad the old chief had retired, he could have had a lot of fun with that one. Had I known this was going to happen, I would have been at the meeting. Taking a day's vacation would have been a small price to pay for being there.

A Train Convention

Not long after this I saw the second most stupid job of running a railroad I have ever seen. I came to work one morning after a derailment had just been cleaned up. Some village idiot had lined up three huge west bounds which did not fit on any siding on the subdivision, to run thirty minutes behind each other. He should have run four trains which did fit instead. Going east he had the much-delayed hotshot that would fit, two huge dead freights that would not fit anyplace, and the sacred cow that was supposed to not be delayed anyplace, for any reason. A logistical nightmare. They were stacked up, one just behind the other.

A dispatcher that I used to work with had a definition for a situation like this. Any time there were too many trains in one area at one time, all delaying each other, he called it a cluster-fuck. As crude a term as this is, it did a nice job of describing such a situation. And this looked like it was going to be the grand-daddy of them all.

The chief I had last worked for in Hogback Hollow had been appointed to what amounted to the super chief's job here. They had given him a nice title, very little extra money, and in the end, a deluxe shafting, in my opinion. He walked into my office, and announced that he wanted me to have the sacred cow pass the other three trains before they reached the next terminal.

My jaw dropped. If I would have been wearing false teeth, I would have lost them. "Please, for the record," I asked him, "How do you propose we get these trains around each other, I don't care in what order?" I knew he had not set this situation up, he was far, far, too good an operating man to even think of this on his worst day. He looked at the train sheet, realized what the situation was, turned ashen, spun on his heel and left, never saying a word. I don't blame him, I would have left too.

Time went by, the three huge West bounds knocked down the signals one after the other in the centralized traffic control area. They were doing a little better than I had expected. The first train was loaded to the hilt, losing a little time on the hills, but not as bad as I feared. The second one was slowly catching up to him. The third train was the longest of them all, but had a large number of empties. He was the lightest and had the quickest engineer.

The largest siding in the area had a branch line running off of it. Checking the train lists, I could tell both the Eastbound hot shots would fit on this siding, but, it would be close. The sacred cow was of course early, just when one did not want him to be. A glimmer of hope. The branch line was just where it needed to be.

I called the trainmaster who served this area, a darn good one, on the radio. He was running alongside his trains on the nearest highway, like the good Shepard he was. He knew the smelly stuff was about to hit the fan, and wanted to be on hand to help when it did.

I told the trainmaster what I intended to do. He did not argue. I called the engineer on the third Westbound, telling him "Hank, rattle them a little, I need you at a meeting point."

Rattle them he did. The train crews, many of them

overhearing the conversation, started chattering like a bunch of magpies. By the time the first Eastbound started into the siding, the crews and the trainmaster knew exactly what they were going to do, and it what order.

Soon both of the huge Eastbound dead freights were on the branch line. Then the sacred cow was in the siding behind the other hot shot. The Westbound monsters rumbled by. The sacred cow backed out of the siding and took off like a scared rabbit, the other hot shot on his heels. Before long both dead freights had backed out onto the main and were rumbling east, now in reverse order.

I spat out more track warrants in about an hour and a half than I sometimes did in a whole day I am sure the trainmaster had more exercise than he needed, acting as flagman, taxi cab driver for the brakemen, switch tender, and cheer leader. I did not want to know how many rules were bent, nor did I care. I would have liked to have hung a medal on the train crews, they could not have done better.

But, by far, the biggest factor had been luck. Pure, simple unadulterated, horseshit luck. Had the branch line been in a different place, or had so much as one wheel gone on the ground when pulling into the rickety branch, it would have been many hours before the trains got moving.

A radio station used the name of this town when they ran a spoof about Dallas, a popular TV show at the time. The oil well there had one barrel of oil in it. The family that owned it was named Kidding; I. M. Kidding, R. U. Kidding, N.O. Kidding, U.R. Kidding, and of course, Mae Bea Kidding.

This town had a new claim to fame. It, for one brief moment in time, had seven trains there at the same time. The mayor in his own private airplane, happened to be in the air at the time, and got a very impressive picture of it. A local paper ran a copy, with the headline "Train Convention in Town." A couple of people promised to mail me a copy of it, but I never did get one. Still, all in all, we had been very, very lucky, and no kidding, either.

A couple of hours later the super chief came into my office. He had a periodic update of which crews were called for what trains at each terminal in his hand. The crew for the sacred cow was called ten minutes early at the next terminal. The train had been a little early to start with, the train light, the engineer quick and able to make up some time. I was sitting there, feet on my desk, having a smoke and a cup of coffee. I needed a break.

"How in hell did you manage this," he asked.

"It was easy," I told him.

"Huh," he grunted.

"Sure," I told him. "And, just for practice, I ran one of those dead freights around the other."

Possibly thinking he had already asked me one more question than he wanted me to answer, he turned to leave. "Oh," I told him, "Please tell whoever lined this up that the Good Lord is noted for taking care of drunks, babies, and idiots. The Lord took very good care of him today." He did not say a word, just left. I had made my point.

Things went along rather smoothly, considering that both the old chiefs were now gone. The air comptrollers seemed to be making good train dispatchers. They were intelligent and had a very good sense of responsibility. They would sometimes laugh at themselves for their instinct to have someone increase their altitude. I think there were six or seven dispatcher's desks in the office. One of them was badly overloaded. Management gave one of the subdivisions they dispatched to our desk.

An Old Spot Revisited

This would be the fifth time I would be working this subdivision, knew it well, knew my own territory well, and decided to remain on my job. A new rule had been put into effect requiring a dispatcher getting a new territory to take a student trip over it. This I did not mind. I would get some time away from the office and get to see some people that I had not seen in many years.

I got a ride to the near end of the subdivision with one of the officials. I got a motel and had the caller notify me when the first crew, after daylight, was called for duty. I don't remember what time it was, but I do remember all the raspberries I got for the next couple of days.

I had been working this territory before some of the trainmen were born and had been working it, off and on, for as long as most of them could remember. I was as old as dirt. They all thought me making a student trip was very funny. I was promptly greeted with comments like, "About time you got out here so you know what is going on," to "Please don't get lost, we don't have time to go looking for you." I think that every crew, lineman, signal maintainer, and section foreman got in a good-natured jab at me.

My sides hurt from laughing by the time I got back on an eastbound train the next day. My wife came to get me. We went out to dinner with one of the crews, spent the night, and went back the next day. It was a nice mini vacation. It had been a long time since I had enjoyed myself more.

The new subdivision was not that much of a hassle for me. We would let the Westbounds run, and call the Eastbounds to fit about the only decent meeting point we had. The trains were long, but there were not that many of them. I did not have to devote that much of my time to it. There were two very bad hills, one at the extreme East end, and the other toward the center. The two young dispatchers who also worked the desk hated it. One could get himself in a world of hurt in a hurry, if he pulled the wrong string.

The last time I had worked the territory, a big Eastbound

stalled on the hill which was in about the middle of the subdivision. The rear end of the train ran away while they were taking the head end into the next town. I was told that it sounded like a tornado when the rear portion of the train went through the town at the foot of the hill. It must have been going in excess of a hundred miles an hour.

Luckily, nothing derailed, nobody was hurt, and the rear end of the train came to rest eventually. After going part way up the next hill, then back down, part way up the first hill, and back down again. The fact it was about four o'clock in the morning when it happened helped.

Time went by, and we got a new trainmaster in this area. He was a nice young guy, but very inexperienced. He watched the operation for a while and became unhappy with it. That I could not blame him for, it was ragtag, but the best we could come up with unless we had some more long sidings.

One afternoon he told me that he was going to run shorter trains, but more of them. None of them small enough to fit much of anywhere and all too heavy for the engines they would have. I tried to talk him out of it. Any rookie dispatcher in the place could have told him it was not going to work. It had no chance. There is little difference if a train does not fit in a siding by fifteen or fifty cars, He still will not fit. Every train would have to double the hill. He went ahead with the doomed plan anyway. I told him he would have the God-awfulest mess he had ever seen by the time I got back the next morning. Nothing would change his mind.

The next morning when I came to work, things were even worse than I expected. One Westbound had been able to scratch his way to where the switch engine crew could go get him. He still sat there because the switch crew had not yet had time to do this. Trains were most anywhere with no crews on them. Two were put away at a station where there was some extra track room, the rest sitting on the main. The new trainmaster had used every crew he had, the first crew, tied up on the line, would be rested toward ten o'clock. It was a real mess. "Jesus, Joe, what should I do?" were his first words to me when I sat down in the chair.

"Is it a nice day down there?" I asked him.

"Sure is," he replied. "Why."

"Do you like to play golf?" I asked him.

"Love to," he replied.

"I think that you have been working too hard lately, and being you asked me, I think that it would be a good day for you to get a couple of rounds in, but please come back tomorrow to sign all the time claims you are going to get," I told him.

To my surprise, he did exactly that. I talked to the crew that was tied up on line, told them what I wanted them to do. They were willing to take about a hundred and fifty cars of grain down that big hill. This put visions of dollar signs dancing in front of their eyes. For the kind of money they would get, they would not have minded taking two hundred cars down the hill.

I got rid of three of the damn eastbound trains this way. The crew ended up with about five days pay, with all the lap-backs, dead heading back and forth, and the dead head home. I don't remember exactly what happened next, except whatever I did cost a lot of money.

The east end chief, whose territory it was, avoided my office like the plague until it was about time for me to go home. I think he preferred not knowing what went on that day. He was pleasantly surprised to see the mess about cleaned up. I told him the trainmaster would probably have writers' cramp from signing all the time claims, he said "Good, next time he might know better." This chief was a right guy. This was maybe the third worst bit of poor railroading I have ever seen.

This new trainmaster learned very quickly. Once he got a little experience he proved to be one of the finest. He was a joy to be around. I wish him nothing but the best.

One day the chief gave me the number of a car that had been badly delayed and now was to be on the head end of the next time freight out of the terminal. He wanted me to make sure it was set out and spotted at an elevator out West. They were out of product. I gave the train crew a message to do exactly that. When they left the terminal I checked with them to be sure they had the message. They had it, and I told them to please be sure it got done.

The next morning the elevator manager called me. He said the car had been set out there, but not spotted. He had to use a truck to get enough product out of the car to supply somebody who was out. I told him that I would get it spotted for him as soon as I possibly could. He was a good customer and deserved some service.

This same crew was now on the sacred cow, on their way home. They were a little ahead of schedule. I asked the conductor why he had not spotted the car, as instructed. He gave me a song and dance about the train he was on being a hot one. He did not want to take the time to do it. He was rather noted for trying to make other people do his work for him. I was sick of it. "I'm afraid you will have to spot it this morning then, I just promised the elevator manager we would," I told him.

"You can't stop this train, this is the hottest train we run," He replied.

"Well, unless you can talk your engineer into running by a red signal, you are stopping. Further more the signal stays red until the car is spotted and you apologize to the elevator manager for the inconvenience you caused him. You told me that the train you were on last night was hot, too, and somebody has to do it."

He quit arguing with me. I called the elevator manager back, told him there would be a train there in about half an hour. Please call me back as soon as the car is spotted and the conductor apologized to him. He told me he would be glad to.

A few minutes later, the engineer on the train called me. "You are actually going to give me a red signal?" he asked.

"No," I told him. "It is red now. It will stay that way until the elevator manager calls me back, telling me his car has been spotted and he got his apology."

I could tell that the engineer was laughing when he said ok to my instructions. I think he thought I was just tormenting his conductor. The chief happened to come into the office while the crew was spotting the car. Looking at the display board, pointing, he asked "Is that the high roller?"

"Sure is," I replied

"What is he doing there, sitting behind a red signal?" The chief asks.

"Doing some way freight work," I told him.

"Why" he asks, as of yet not looking at me like I had taken leave of my senses.

"Disciplinary action," I told him. He was smiling now, he knew me well enough to know there had to be a story there. I told him about the car not getting spotted last night. This conductor was good at sloughing his work off on somebody else. I was doing this to teach him a lesson. I pointed out there was a Westbound that was going to delay him at the meeting point otherwise, the elevator manager was to call me momentarily, and the train should get in on time.

My phone soon rang. The signal turned green, the crews that had heard my radio conversation with the train started needling the conductor. They called him the king of way freights, tormenting him any way they could.

I would imagine they were sick of doing the work he neglected, and rubbed it in pretty hard. Even his own crew heckled him. The chief left, laughing. The meeting point worked out to perfection. The train was five or ten minutes early out of the next terminal, so all was well.

When it was almost time to go home, the super chief stuck his head in the door. He asked me to please stop in his office on the way home. He had apparently heard about the incident. I no more than stuck my head in the door when he asked me, "What is your definition of a hot train?"

"One without brakes," I replied.

"That's about what I thought you would say, now get out of here, you son of a bitch." He was growling but there was laughter in his eyes. He had made his point, but I laughed a little too.

An Old Friend Shows Up

It was about this time I heard the operator that missed a train order when I first started dispatching was now appointed the acting general manager of the railroad. I knew that he had been with the acquiring road for several years. He had risen through the ranks very quickly. I would imagine that very deservedly so.

I was very glad for him. I had not seen him in a lot of years. We had talked on the telephone a few times. When we got a new trainmaster and company policeman at Hogback Hollow, both much larger than me, he would call and ask "How is your ego doing, now that you are no longer the biggest guy around there?" He loved to tease me, any way that he could, laughing all the while. One day I told him I was so used to the situation by now I felt just like a dime between two nickels. He really got a kick out of that one.

Later, more to get away from rush hour traffic than anything, I bid in a temporary night chief's position. It would be open for several months, the woman who held the position was on maternity leave. Now I had to deal with the assistant, whatever his title now was. I will not bore the readers with all the details of how kind I was to him, but as an example, one evening he wanted to bootleg an empty car on the rear end of a unit potash train. I at first refused, but relented if he would send me a message saying what he wanted, with the car initial and number, for my records. The message was soon on my desk.

The car was picked up on the rear of the train, exactly as he asked. There was a blizzard blowing in North Dakota. The trains behind this one already had engine trouble on their lead units, the snow causing the traction motors on them to get wet. The engines on this potash train were just what I needed to get the hot shots over the road. The train was put away in the yard at the first terminal and the power appropriated. This left the empty on the wrong end of the yard, soon to be snowed in about eye ball deep to a tall elephant.

I splashed the fact this car was there all over the night letter that I left for my relief. Copies of it, of course, went to the powers that be. I did my level best, it would seem, to prevent delay to the car, per orders of the former assistant. I knew full well that digging this car out of the wrong end of a yard in a snow storm would not be high on anyone's priority list, but it gave me a good excuse to inform everyone of the fact that it was there. It might take a week before anybody dug the car out. I knew some people would take big exception to bootlegging cars. He had provided enough evidence to hang himself. I was going on vacation the next day, fishing to Mexico. I caught a nice blue marlin. I also think I know what he caught.

One of the reasons given for the consolidation of all the offices was all of the overtime they could save. Bean counters may be good at counting beans, but not train dispatchers. Some of the dispatchers went back to being operators or clerks, some took early retirement, and some went to work for other railroads.

I do not know how many good, reliable, experienced

people we lost, but it was quite a number. There was the renewed hassle of hiring and training new people, which, in maybe two years time, were still not as good as the people they had replaced. Overtime costs were supposed to drop, but instead went up 35 percent.

It was the same with the people who called the train crews to work. Here, it was even worse, to top things off our callers now did not know the crews they were calling. They did not have the faintest idea of where to find someone who did not answer his telephone. When the calling was done on a local basis, the caller knew the men, usually able to come up with someone.

Maybe, since Christmas was coming he was helping his brother cut Christmas trees, or maybe he was at the Elks playing cards, or whatever, they had an idea where to find a man. Now they lived maybe a thousand miles away. A caller did not even have a clue to where the plow for Adam's off ox was stored, much less where a crew man was. Trains now sat for lack of man power when they should not have.

I have a sneaking suspicion there was a little "I'll lay off to give you a couple days overtime, and you do the same for me," going on. I did not know, nor care to know. I do know that it used to be most any dispatcher had to be very sick or had to be someplace very important before he would lay off. He did not want to cause any hardship to the chief or the other dispatchers in the office. Now, many of the dispatchers did not even know each other, so who cared about anybody else. They were more prone to lay off whenever they wanted. Anyway, management, again or yet, had a tizzy about all the overtime.

The super chief put out a notice that, effective at once, no one was to lay off without proper authority. They would be sent to the company physician if they said they were sick. This notice was not going to affect anything I would do, so I paid scant notice to it. One night about ten or so, the young gal who was working one of the Hogback Hollow subdivisions came to my office. Her office was not part of my territory, so I thought it was rather unusual. Looking at her a little closer, she looked like she had been crying. She was about six months pregnant and I could tell there was something wrong. "Joe," she said "I need a day off."

"Why, what's wrong?" I asked.

"Nothing, really," she said, tears again starting to form in her eyes, "I just need some rest."

In the state she was in, I did not even want her back in a train dispatcher's chair. "Come on," I told her, "I'll do better than that for you, go back to your office and get your stuff. I'll take over for you, you take the rest of the day off, tomorrow, your regular days off, and we will see how you do then." She gave me a big smile and nodded.

When we got back to her office, she had second thoughts. "You're not qualified on this desk," she told me. I laughingly told her I had worked this position for about eleven years without getting into too much trouble, and could work it for a couple more hours without getting into too much more. I asked her if she was ok to drive. She assured me she was, and left.

She had issued some warrants just long enough to keep the trains moving, not having made up her mind where they were to meet. My first order of business was to change this as I did not want them all calling in at the same time.

The first engineer I called got a big F in railroad radio procedures, but an A+ in voice recognition. "Keeerist, when did they dig up the graveyard?" he asked.

Another of them responded, "I heard that they were scraping the bottom of the barrel up there, but I didn't know things were THAT bad."

It had been twelve to fourteen years since I had talked to most of them. I could tell they still loved me. It's just some people do not have any respect for their elders.

I rather enjoyed my brief visit to Hogback Hollow. I finished up some of the chief's work as I went and some after I had been relived. I made arrangements to relieve her position and went home. The next day the phone rang. It was the super chief asking that I come in a little early and talk to him. I always came in early, anyway, so that was no problem. He did not look at all happy when I got there. "I sent that gal to see the company doctor," were the first words out of his mouth.

"I think it might have been better to let her get a little rest first," I told him.

"She took a day off without authorization," He told me.

"Well, I authorized her to be off, don't blame her, blame me," I told him.

"What makes you think you had the authority to let her lay off?" he asked.

"Nothing I have seen says I don't, I was the closest thing to authority she saw, so she came to me," I told him. "What do you want, to be called in the middle of the night for every petty thing?" I asked him.

I am pretty sure he had heard the story of the "West Coast Welcome" Lump-lump got, so that was as far as he wanted to go down that path. "Why did you let her off?" he asked.

"She looked like she needed it," I replied.

"What do you think you are some kind of a doctor?" he asked.

"No, and I'm no vet either, but I know the difference between a live horse and a dead one, and this one looked like it was about ready to drop. In fact I asked her if she was ok to drive, sent her home, and sat in for her a while. I didn't want her in the chair in the condition she was in."

This seemed to displease him even more. "You're not qualified on that desk," he told me.

I laughed. "I've worked it longer than the three dispatchers you have in there now combined. I didn't think there would be too much of a chance I would destroy it in a couple hours," I told him.

"What do I say if someone finds out?" he asks.

"I don't care what you tell them, tell them I was breaking in there, you don't know, or to go ask me." I said. I could tell he was still unwilling to drop the subject.

Just then the phone rang, and he hit the speaker phone button. It was the nurse at the company physician's office. They informed him they had pulled the gal out of service for ninety days. I got up, shrugged my shoulders, gave him a big smile, and walked out. He never said another word. The conversation had been a little tense, but cordial enough.

I felt sorry for the man. He must live in constant fear of higher authority. That would be a hard row to hoe. I am getting too old for this kind of shit, I thought to myself. Every other chief I had ever worked for would have given me a slap on the back and a big thank you under these circumstances. Retirement looked better and better. I could almost reach out and touch it.

One day the phone rang. To my surprise it was one of the BN chief dispatchers that had once been one of my students. "I have my tail in a crack," he told me. "I had a new dispatcher working. He forgot to give your people the twenty-four-hour notice that we have to give you on a coal train. The next dispatcher thought it had been done, and here I sit, with my thumb in my ass. I called, everybody was sympathetic, but nobody cares enough to do anything about it. I have trains out there as thick as the hair on a dog's back, and I can't afford to tie up one of my mains to store the train on. Can you help me?" he asks. "I don't care what it costs, we'll pay any extra expenses involved." I told him to call me back in about an hour, I would sure try.

Looking at the callers' board I found an engineer who was on his days off. He was building a retirement home not too far from where we got the coal train. The caller called him on his cell phone. He was indeed there, working on his home.

I told him I would make it worth his while if he would go. If nothing else, he and the trainmaster could pull the train onto our track. We would have no conflicting trains. Vision of dollar signs danced in his head. Sure he would go.

There was a conductor from the acquiring road, even closer to retirement, who already lived there. He was also rested. Yes, his wife said, he was on that end of the railroad, but out fishing. Soon the trainmaster, who had hired the resort owner to take him out on the lake to find the guy, was calling on his cell phone. Sure he would go, but he did not have a key that fit a BN caboose. I told him I would do my best to find him one, but if I could not to bring a pry bar and Jimmy the door open, use a rock, or even a can opener if he had to.

Now I needed a brakeman. There was a work train on duty about 30 miles away. I got lucky, the train was working in the yard, and the conductor was in the depot doing some book work. It was Friday, and the work train would not be working the next day. I asked the conductor if he minded if I stole his brakeman. "Hell, no," he said. "Fact of the matter is that I would pay to get rid of the lazy son of a bitch. We won't be going anyplace, anyway, just working here. I don't need him."

"Fine," I told him. "Get his ass in a cab, we have a coal train coming." He assured me he would do so immediately.

We had a dispatcher working in the office who was all business. A faint smile at work was about the closest he came to any humor, and we seldom saw even one. He never used a foul word, both looked and acted as professional as they came. If one wanted to do a documentary on train dispatchers he would have been ideal to star in the lead role.

I went to his office, and asked him to check with one of the agents on his line to see if they had a key for a BN caboose. Someone was talking on his dispatcher's telephone and he called the agent on the carrier phone.

"Do you have a BN Caboose key there?" The dispatcher asked.

The agent, not recognizing his voice on the speaker phone, thought it was someone else and replied, "I am sick of your God Damned Polish Jokes," and hung up.

The poor dispatcher looked at the receiver like it was a cobra that had just bit him.

The crew was waiting when the coal train got there. The conductor and engineer got about four days pay to go someplace they had to be the next day anyway and stayed at the motel the company provided. They were happy campers. The brakeman got some dead head, a couple extra days pay, and ended up where he was going, so he was happy. The trainmaster found a BN caboose key in the glove box of his company vehicle, so the caboose was saved the indignity of being broken into. The BN chief was very happy, and said that he owed me one.

The task did not take me very long to complete. The next time the assistant needed a jab in the posterior, I could maybe call in my chip that the BN chief had given me and I might be happy too. I don't suppose the assistant would be too happy when the BN chief applied his prod. He would not be too gentle, but somehow I did not care about that. I never did call my marker in. The BN chief was a good guy, I was glad to help him.

The papoose was born, and I went back to my own job. One day the former assistant was in town and told me that next week the general manager would be here. He was going to have a talk with him about me. I gave him a smile and told him, "That should be interesting."

The next week I had delivered some information to the chief's office and shot the breeze with him for a few moments. As I came out the door, I could see the former assistant standing there with his boss. Just then the hallway door opened and in walked the general manager. We saw each other at about the same time. He charged on past the assistant and his boss, heading straight at me.

We had not seen each other in a good number of years. We made enough noise that heads were stuck out of office doors, much like a bunch of gophers looking to see if the cat were indeed gone. After we had done enough bear-hugging, back slapping, hooting, and exchanged enough "How in the hell are yous," we decided we had both better go back to work. I did not lose much sleep that night, wondering what the assistant was going to tell this man about me.

Sometime later, I spotted the former assistant in the outer office. "Say," I asked him, "You wouldn't happen to know when the general manager will be back in town again?"

"Why?" he asks.

"I just thought it might be interesting to talk to him again, that's all." I gave him a big smile. He never said a word, just stood there, sort of green around the gills. There was no way I would have bothered the general manager with something like this, but he did not know that. Let him wonder and stew, it would give him something to do, and he would not bother so many people. After this, I only talked to him three or four times, which was more than enough. He was courteous to me and I to him. I had made my point. And I never hung him out to dry again, either. I had no need to.

Back To Hogback Hollow

The guy working one of the Hogback Hollow jobs retired. I bid the job in. It was the best one in the place. The super chief was on vacation. The chief asked me if I wanted to make a student trip. He said that he would not force me to go if I did not want to. I declined. My back now hurt, it was hot, and I had no desire to go mucking around on a dirty engine for three or four days. He stuck someone into the office to break me in. That worked out well, we could each work for a couple hours and than take a nice nap.

The yardmaster of a neighboring subdivision let a train out of his yard loaded seven hundred or so ton over the tonnage ratings for the engines he had. We had a bad hill just beyond the next terminal. I saw who the engineer was and decided it would take about as long to reduce the train as what he would lose on the hill. I let him go with it. When I thought he was about to the top of the hill I called him on the radio, asking him where he was. "Where do you think I am?" was his response.

"With that monster, I would think that your engines are about half way between the siding switches, right on top of the hill," I replied.

"That's just exactly right," the engineer replied in a very angry voice.

Thinking I could flatter him a little, I said, "You are the best there is taking a big, heavy train up a hill."

"Ya and you are a son of a bitch," He replied

"Well, I've been called worse, but what makes you say that?" I asked him with laughter in my voice.

"Well, you know that I can do it, so you give me the cars," He replied, still angry. Evidently his anger was caused by the fact I had not overlooked the size of the train, since I knew exactly where he should be.

"Well, I gave you the biggest compliment as an engineer I could give you when I sent you out with that train, even if I am a son of a bitch." I told him. This seemed to cool him down.

Two or three weeks later, the caller advised me there was no engineer available at Hogback Hollow to go East. I had a preference train en route. A damn nice one it was.

The caller's office was right across the hallway. "How is my favorite call girl?" I asked her, when I got there. She sat there patiently. She was looking at me as if now thinking you have had a little fun, what do you want, you old buzzard. There was no way on God's green earth I was going to let this train sit, if I could help it. "Who do you have, off the West end that is rested?" I asked her.

"Just this one," she replied, pointing to the name of the engineer who had called me a son of a bitch, "and I already asked him, he says he is not qualified."

The man had spent about twenty years running back and forth over this pike, but she had no way of knowing that. "Call him please and let me talk to him."

She calls, he answers, and I tell the man, "I'm going to have a hell of a nice train sitting soon if I can't find an engineer. I know if you don't want to go, you don't have to, and no hard feelings if you don't go, but please don't give me any happy horseshit about not being qualified, or I might call you a son of a bitch. On my railroad you are qualified to run an engine, even if it were on the dark side of the moon I'll dead head you right home. Would you go as a personal favor to me?"

He never even hesitated. "For you, Joe, sure." He couldn't have been too mad at me. He was one of the best. The caller did not seem too happy with me, finding a man when she could not. I felt bad about that. The favorite call girl bit was not just all fun, she was my favorite caller, and always did a nice job.

The callers did not always have such an easy life, either. The chief caller, who was promoted to this position just after she qualified to become a train dispatcher, suffered a nervous breakdown shortly thereafter. It seems that the carrier was putting undue pressure on her to find men when none were available. From what I hear the former assistant was responsible for some of the pressure.

I don't know if they expected her to give birth to more crews, or what they expected. I think she was also exposed to pressure regarding all of the overtime her callers got. It is too bad.

I often wonder how she would have done as a dispatcher. Life is strange. She had been a very good student, very intelligent, conscientious, and had passed her rules test easily. There was every reason to think she would become an excellent dispatcher. Something that feels like a lot of pressure to one person can be sloughed off by another as just more noise.

I guess this woman did not have the ability or the desire to say, "tough, go hire some," when somebody got on her case about the lack of men. I also have a hunch she took her problems home with her, something that she would not have to do as a dispatcher. This is the reason that I never particularly liked working a chief's job, as a dispatcher you tend to leave any problems behind when you go out the door. Either the chief caller or the chief dispatcher is called at home with everyone's problems. She did not need this either.

I had a problem with an extra gang foreman. It was a small gang, but the foreman was a big pain. He did not believe anybody's line-up figures, and would delay every train. He was working in centralized traffic control territory. He would shunt the track, using a crow bar or something to short the circuit out so that one could not line a signal through where he was working. I chewed on him, the road master chewed on him, the chief sent him nasty messages, but all to no avail. He slowly worked his way East, into track warrant territory.

The gang foreman came in shortly after lunch one day. He asked for a better figure on an Eastbound that I had on the line-up for four that afternoon at the point where the track warrant territory started. I told him the train would be there at four, he was just leaving the terminal. He laughed at me. "There are some slow orders on the Eastbound, he has trains to meet, there is no way in hell he can be here at four." I told him four was it, to get out of the way. He laughed at me again.

I knew a few things he didn't. The engineer was the

quickest I had. His train was light and could go up the hill on the westbound as quick as he could go down. Maybe somebody would get delayed at a meeting point, was it was not going to be this train.

I called the road master, asked him what the gang was doing and what kind of equipment they had out there. Just hand tools, he told me, he was to pick them up at a crossing at four thirty to take them back into town.

I told him that this train would be right on the line-up figure, he was out there with a company truck, and I did not want to get that hit, it could run on the rail. I called the engineer, told I wanted him to be as close to the line-up figure as he could be. “Keep them high and green, I’ll be there,” he told me. I gave him a track warrant at once, and thanked him.

I was my habit to give an Eastbound train a warrant at the station West of where the track warrant territory started, and everybody knew it, including the gang foreman. I figured he would have his portable radio on, listening for it. When he heard nothing, he kept on working.

There was a curve and a grove of evergreens just West of where the gang was working, with their kegs of track spikes, bolts and nuts, lunchboxes and coolers all neatly stacked on the crossing waiting to be picked up. The engineer, maybe more ticked at the gang foreman than I was, shut his engines down to idle before he rounded the curve.

The wind was blowing in the right direction, making it impossible for the gang to hear the train until the engineer blasted his whistle when he was only a few yards from them at about 4:01pm. All they had time to do was dive for the ditch.

I think I would have rather paid their laundry bill that night than washed their underwear. The train hit the stack of stuff going about sixty miles an hour, scattering it all over the landscape. The engineer told me later that it sounded like the worst hail storm in history when the big bolts and track spikes came down on the roof of the engine.

This road master and I had worked together before. He was a wise old bird, and knew exactly what was going to happen when I asked him where the equipment was, and warned him the train would be right on the figure. He managed to hide someplace and drive up to the spot just as the train got there. I imagine that he gave the foreman a good chewing, and then made them pick up the track spikes, nuts and bolts, on their own time. It was June, and they were out there on their hands and knees until after dusk. He had his own way of teaching them a lesson.

When I was at Hogback Hollow we had a big heavy load derail in a sag just the other side of the crest of the big hill coming into town. It chewed the ties up very badly, there were no ties available to fix the spot and the section foreman placed a 5 mph slow order on it.

There is no good spot anywhere for such an order, but this was the worst. The next morning I told the section foreman that we had to get that spot fixed before we put something else on the ground. The section foreman said he knew that, had no ties to do it with, even if the spot was small. He looked like he was about ready to cry.

I saw there was a load of ties on an incoming train. I told the yardmaster to put it someplace that the section men could get at while he switched the train and then put the car back into the train.

When I told the section foreman to unload the top row of ties, use what he needed, and hide the rest in the weeds, he was very hesitant. He was as honest as the day was long and not sure he wanted any part of this. Besides, he said, the road master would be coming through next week on an inspection trip. What would he tell him when he saw the new ties? The road master knew he had none.

"I'll take care of the Road master," I told him. "Just send

him to me if he gives you any static. If it makes you feel any better, put a push car or two of gravel over the area, and maybe nobody will notice them."

A while later the section foreman comes back into the office all excited. He tells me that these were some special ties, heavier than normal, some kind of bridge tie he thought, and he dare not steal them. I told him "Good, if they are so special, you can have two rows."

I don't think he took the second row, but that afternoon the spot was fixed. The following week he went out on the inspection trip with the road master. Someone brought him back from the end of his territory. He walks into the office. I ask him, "How did the inspection trip go?"

"Well," he said, "We stopped at that spot, he got out, pushed some of the gravel out of the way and told me. 'Hey, them is nice ones. That Joe is taking right good care of you, isn't he?'"

We had a flood on the West end. A bad one, washouts everywhere. The road master, his assistant, and a couple of section foreman were out measuring them, reporting the length and depth of each to me on the phone. Using a formula the superintendent had taught me, I was figuring out how much rock we would need. I had little else to do. All I had to worry about were a couple of locals and the branch line train. About noon the superintendent calls, and there is some bean counter on the phone with him. He asked me if I had the figures yet. "Just got them done a few minutes ago, we will need about two hundred and sixty-five cars of rock." I told him.

The bean counter said that he did not think that this was possible, to let the boys with the slide rules have a crack at it.

"Hell no," says the superintendent. "If Joe says we need that many, that is what I want, but send ten more, just in case. And get at it right now," he adds. He knew darn well that if he would have been presented with the same set of figures, he would have come up with the same answer that I had.

In due time the cars were unloaded, and the road master's official count was two hundred and seventy-one cars used. I was talking to the road master a couple of days later. "You are a brown-nosing rascal," I told him.

"Why?" he asks.

"Well, I padded my figures on that rock a little, the boss ten more, and I know darn good and well you did not use that much rock. You just want to make the boss look good since his figure was closer than mine," I told him.

"Well, I just might have hid a few of them somewhere," he tells me.

"Yes, like on a spur track behind an abandoned old roundhouse that I happen to know about. Where did a nice honest guy like you learn such a low-down, filthy, dishonest trick like that?" I asked him.

"You go to hell," he replied, hanging up, his laughter ringing in my ear. It is always nice to know the people you deal with well and also to know where there are a few cars of rock in case you need them in a hurry.

I was now working with a new trainmaster. We had been telephone friends for about fifteen years before we ever talked in person. We had worked together at hog hollow, briefly for each other, and ended up in the same office after the consolidation. He had been promoted, and now had the territory I worked. That, in fact, was one of the reasons I went on that job. I talked to him almost every day, sometimes several times, and not always about business. We never had one harsh word.

When the trainmaster was in town, he usually stopped to see me. If I got the chance I would stop to see him. The better I got to know him, the more I thought of him and admired him. He was slow to anger, quick to laugh, and thought things out well before he acted. Besides his obvious intelligence, I think his biggest attribute was his willingness to listen to what people would tell him, be they the president of the railroad or a track hand. Not that he always agreed with them. I would say that he was the perfect example of the old adage. "A wise man learns more from a fool than a fool learns from a wise man".

I vowed to myself to keep in contact with him after I retired, and that I would keep him as a friend as long as he would have me as one. He did not disappoint me. He was one of the best.

The last sharp look I got from anyone on the railroad was from the super chief. He had put out a notice regarding a dress code for the dispatchers. There was nothing in the union contract about a dress code. He should have gone through proper channels, but did not.

One of the rules he had was nobody was to come to work in a shirt without a collar. My wife had purchased a very classy shirt for me that had no collar. I had her lay out a matching outfit for me, sports coat and all. Everything all color coordinated. She is good at that sort of thing, The morning the notice became effective, shoes shined, looking as good as I ever could, went to work. I would feel very comfortable giving a bridge lecture on a cruise ship in this outfit. In fact, I wore it several times doing exactly that.

I came in the door finding the chief doing an inspection of his troops. He saw me and did not quite know what to think. Then he noticed my shirt had no collar. I got the sharp look, but he did not say a word. Not too much he could say, I was undoubtedly the best dressed man in the place. I think that I had made my point. It was not what he did, it was how he did it that I found offensive.

It was the only day of a thirty some year dispatching career I wore a shirt with no collar to work. I knew that the notice was not directed at me, and had no problem with it. I had started my dispatching career when a suit and tie had been the standard attire, and had not digressed that much. I had often found it a pity that some of the people now showed up like they came to clean the sewer instead of dispatch trains. It is strange what a difference a generation can make.

There was also another difference. When I started, the dispatcher was considered as someone who was, more or less, the cream of the crop. Possessing a skill few had mastered. A dispatcher drew immediate respect from most everybody on the railroad. Now we had mostly people who, without the use of a computer, could not survive one hour in a dispatcher's chair without dire consequences. We had started out as the elite, and now ended up, too often, as a pissing post for everybody on the railroad. Maybe this is why some of them dressed like they were there to clean the sewer; they were all smarter than me.

On the other side of this coin, a few years before this, a young dispatcher and I were called as witnesses to a standard investigation. I showed up in a passable suit and tie. He showed up looking his last meal had maybe come out of a dumpster. After the investigation the young dispatcher was angry. "How come," he asked, "Do they seem to treat you with so much respect, and me like a piece of shit, just because you show up in a damn suit?"

I told him I did not see much difference in the way we were treated. If there was any it was that he was still somewhat of a kid and I was an old fart. He did not believe me. Outside of maybe some long looks at his attire, I had seen no difference what so ever in how we were treated. Both well, as far as I was concerned. I think the only difference is he felt out of place there. He looked for and found things to take exception to.

The general manager did not show up at the office very often, and when he did, he was always in a hurry. He was prone to stick his head into my office, shake his fist or thumb his nose at me, laughing all the while, but seldom stopping. Beneath the expensive suit and heavy responsibility he bore on his shoulders, there still lived this mischievous boy I was so fond of

Then one day he walked into my office with no sign of this boy visible. This was no social call. "How well do you know this guy?" he asked, referring to the former assistant.

"More than well enough to suit me," I replied.

"What do you think of him?" The general manager asked.

I had not lied to the man in the thirty or so years that I had known him and was not about to start. "My mommy told me if I can't say something nice about someone, not to say anything" I replied.

"That's about what I thought. I have nothing but complaints about him. Well, OK." He got up off the edge of my desk, said that he heard I might be retiring soon, shook my hand, and wished me well. The boy showing his face once more. We chatted for a moment, and he was gone.

I saw the man only once after that, and only for as long as it took him to shake his fist at me. Another social call.

The railroad was acquired by another, larger one. I understand that the general manager later got into a spat with someone and went to another railroad. I heard various versions of what happened. May he be in heaven ten minutes before the devil knows that he is dead.

Not long after I had retired, I heard the former assistant was told to seek employment elsewhere. I also heard that he and a coworker got into a fight and they both got fired. I feel sorry for the other guy; I always liked him. As for the assistant, the devil may not want him, he maybe couldn't stand him either.

The super chief also ended up seeking employment elsewhere. I was not there to see it but I have a very strong hunch his inability to stand up to senior management was his undoing. Instead of sticking his feet in the sand, screaming at them that something would not work, he went along with them.

Then, needing a scapegoat for all that went wrong, they blamed him for something he could not make work and should not have been tried in the first place.

When they gave him the super chief job, he withdrew from the union and gave up his seniority. If this was done voluntarily or not, I do not know. I do not imagine that it took him over ten minutes to get a job elsewhere as a dispatcher. But, who wants to look forward to being the youngest dispatcher on the roster at his age. I think they shafted themselves worse than they did him. He would be almost impossible to replace. He was a very good operating man. Senior management could have easily given him some sort of job doing exactly this. He had his faults, but I do not think he deserved this.

The East end chief was given the job. I imagine he did much better standing up for himself. He would be the last of the old time mother hen chiefs. He tried to take care of the men the best he could under the circumstances. He is a right guy. I heard he recently retired and his position was filled by someone who had never even been a train dispatcher. The queen is dead. Long live the queen.

The End Or A New Beginning

I was now having rotor cuff problems, having reached for one too many levers to throw a switch or a signal. I had missed some work because of it and now could work only three days a week. Then I was told that the condition would never get better as long as I continued working. My back hurt. I had enough. I told the chief when my last day of work would be. Five or six years after I retired a doctor discovered that I had a compression fracture of the lower back when I was very young and bone spurs had formed.

When I started work as a young kid we used an iron telegraph wire and steam power. I was now a dinosaur, the last dispatcher from either of the two railroads with steam experience.

Now we used a computer instead of brain power. The computer, at least most of the time, kept you from doing anything dangerous, but it did not keep you from doing something stupid. When we first started using a computer someone asked me how I liked using it. My reply was that it was something like having sex wearing a dozen condoms, maybe safe enough, but not very satisfying.

Shortly before I retired, they revised the track warrant form. They added a line that let you run a train, at restricted speed, through an area occupied by other trains and engines without specifying which or how many trains or engines were in the area. I refused to use it, as did some of the other old heads, using the line that did specify them. The rules department questioned this; it was faster using the other line. I told them I had engineers out there, not kamikaze pilots. We begged them to remove the line. They did not.

A year or two after I retired a train using a warrant containing this fatal line came through an area that usually had a switch engine and a local in these limits at that time of day. Both the local and the engine told the train they were in the clear. There was a siding just to the west of them. A train whose crew had run out of time and was no longer on the train was sitting on the main line, the relief crew not having yet arrived.

There was a curve, the weeds possibly hiding the rear end device on the rear car. The engineer, thinking at first these cars were on the siding, was unable to stop before hitting them. If there had been a caboose on the train the markers would have been plainly visible. He would have known at once that it was a train, very possibly on the main track, and could have easily stopped. He did not hit it all that hard, but hard enough to throw the rear car of the train, an empty tank, up in the air far enough that it came down on the cab of the engine, killing both men in it.

Did the computer let the dispatcher do anything illegal? No, to my knowledge he observed every required rule and regulation. Dangerous? Usually, but not that time. Stupid? Always. All to save a moment or two of the train dispatcher's time.

The engineer was a friend of mine. Would he have struck the train in the rear, if someone had spent ten seconds telling him this train was out there? Hardly. He was among the finest of engineers that I ever worked with, alert as they get.

One of the most unpleasant things a train dispatcher faces is a crossing accident. There has never been an automobile or a truck built that has ever won an argument with a train. Having people killed while on duty is not a pleasant thought to go home with.

My personal worst was having ten people killed in an eight-hour shift. Two were single fatality incidents. The other was a Volkswagen mini bus with eight teenagers in it striking a grain train in the fog. The train drug it and rolled it for about a mile before it broke the train line, causing the train to stop. What had been a van was now a huge ball of metal, the remains of all eight inside it.

I can only imagine what an engineer feels like, sitting there, unable to even come close to stopping, when someone darts in his way. It must be a sickening feeling. His worst nightmare is a truck hauling gasoline and the fire that is almost sure to follow.

I remember a case when a train struck one. Both the engineer and brakeman were able to get out, clothing on fire. After rolling around enough to extinguish his burning clothing, the engineer shooed his brakeman to safety. Putting his life in grave peril, he then cut his train and remaining engine away from the one that was on fire, and backed his train to safety.

That is the kind of a man and employee that he was. There

should have about thirty- five-hundred gallons of fuel in the tanks on the burning engine. I will leave it up to the reader to imagine what happens when a fire gets to this much fuel. The bean counters evidently thought this kind of a person was expendable. He was the one rewarded with the kamikaze track warrant that took his life. I could still cry.

One very positive thing the railroad did was institute something that was known as "Operation Lifesaver." It did much to educate the public to the fact that it took a train the same amount of time to go over a crossing, whether their car was on it or not. I am sure that this saved many lives. I happen to know one of the engineers that was selected to be a part of this program. They could not have selected a better man for the job. I am sure that he had much to do with the success of the program. May the Lord bless him for his efforts. I would never let anybody tease this man, that is my job, and I have the seniority to hold it.

There is a long-running debate about the use of a caboose. They are sort of a pain in the tail, requiring personnel to clean them, to fuel the stove and provide drinking water. The result of using one is being able to handle one less empty on the train. Sometimes the caboose was used by the crew as a Pullman berth, sleeping en route.

Yet, I feel the caboose is worthwhile. If the crew is doing its job, it can prevent very expensive derailments by watching the train and the track behind it for fresh marks indicating there is a car derailed, stopping the train before a major derailment happens. It saves backing up when a train pulls out of a siding to pick up the man that closed the switch behind it.

Having someone on the rear of the train when there is trouble, say a bad air hose, can save a lot of delay. I understand they are now going to a two-man crew in some places, leaving nobody to ride back there if they did have a caboose. Yet, I feel, from a safety standpoint, the caboose is still king. Is there a bean counter that can put a price on a human life? He might count more carefully, if the life in question were his own.

I was around the railroad for a long time, and could never understand why they tried to save money by overloading the train dispatcher. It makes about as much economic sense to me as buying a very expensive Mercedes Benz automobile, then buying a cheap ignition switch that causes it to miss and sputter, when one for a buck more would make it run fine, or at least much better.

I also could never understand why they chose to move all the offices so far away from the territory the men worked. When I started, the chief dispatcher and the superintendent ran each division without much outside interference, knew the territory and the men well. Now it seems that hardly anybody knows any territory well, can only slap a band aid on it when it bleeds, and roll with the punches. From what I hear, this has gotten even worse since I retired.

Another thing strange to me is that I have worked for some very good chiefs over the years. I consider three of them geniuses, yet not one of them was retained in any capacity after he retired. They will take their vast store of knowledge to the grave with them, none of it passed on to others after the moment of retirement. Retained as some sort of a consultant, they could have imparted a vast amount of knowledge to many. I feel this was a huge loss to the railroad industry.

If the average person walked by a train dispatcher's office and saw the dispatcher with his feet on the desk he would conclude the man was idle. If he saw him putting out train orders or track warrants, he would conclude the man was very busy. Almost the reverse is often true.

Putting out the orders or warrants is easy, a trained ape could almost do it. The hard part was accomplished while he had his feet on the desk, figuring out exactly what he was going to do. A good train dispatcher's brain never has its feet on the desk, it has too many things to be concerned about.

Looking back over the years, I feel I was a very fortunate fellow indeed. It was my good fortune to work for all three of these geniuses. With the exception of one man, I feel that I parted on friendly terms with all the men I worked with and for. This is a good feeling.

It is also a good feeling that I, nor anybody I ever defended as a union official, except for the man who was drunk on duty and could have expected a six month or year suspension, ever served a day off without pay.

It is an even better feeling to know I managed to conclude my career without ever causing any damage to company equipment through any action I took.

The best feeling of all is that I managed to get through all these years without anyone getting injured or killed because of me. Here I feel that I was extremely lucky. Being careful and fairly competent is wonderful, but lucky is much better.

Maybe part of this luck was due to the fact I always tried to be as good as I could to the train crews and operators. The operators were the eyes and ears of the dispatcher. I had some good ones.

I think the train crews watched out for me as much as I for them. As I noted at a fairly early age, the lead in my pencil far outlasted the eraser, so I tried to be tolerant of the mistakes they made. As a result, my carpet had some very large lumps under it, and many of them.

If I had a problem with someone, I would take it up with them whenever possible, not management. This fact the few malingerers we had discovered in a hurry. I owe a very large thank you to all the people that helped me so much.

One thing I always felt was in my favor was that my personality is not one to seek change. When something has been done in a certain way for many years, the odds are good there are many reasons for this. Someplace along the line someone already tried to change it without much luck.

If someone asked me if something would work or not I always told them the truth as I saw it. If it did not work, they could not be mad at me. If it did work, it cost nothing to say I was wrong, and that I was happy it worked out for them. It was a win-win situation. They were happy with me either way. I felt that I was being paid for getting the trains over the railroad in a safe and expedient manner. This I always tried to do.

There was nothing remarkable at all about that last day. Nothing even hinting of the unusual took place. I cannot bring back any of it except the final minutes. It was now close to quitting time; I had one final train left to fix. This would be the last authority for a train to move I would ever issue.

My mind wondered to how far trains had moved, using the authority I had issued to them, be it by warrant, train order, or signal. If one had ten trains moving at forty miles an hour for eight hours, they would make over three thousand miles. Working 50 weeks a year for thirty-two years, one would come up with twenty-four million or so miles. Roughly a thousand times around the earth. I am sure I did not average that, but still the total would be very high. Trains do not fly, but they don't stand still either.

I got the conductor on the radio. I told him after this he would not have to put up with me any more. This was the last time I was going to do this. He in fact saved the warrant and gave it to me a couple of years later.

His was to be the only train on that part of the subdivision when the warrant had been issued. I glanced to see if any Eastbound would be a conflict. Nope, there was one coming, but not soon enough to get out of the terminal to meet him. I gave him the whole God damned railroad. He was going home and so was I.

The end

Definitions

Air - Refers to the braking system on a train, all cars are connected by an Air hose Pumping up the air is recharging the system

Bug - a device used by the telegraph operator when he sends Morse code it makes dots automatically

Caboose - A car hauled on the rear end of a train. It was equipped with a desk for the conductors use, had bay windows and a cupola. It was stocked with flagging equipment and drinking water.

Crew change point - Place where one train crew replaced another.

Dynamiter - A car with a defect in it's braking system that results in an emergency Application of all the brakes any time the brakes are used

Extra gang - Group of track laborers temporarily assigned to repair the railroad in a given area.

Fill - Cars added to a train to get it up closer to the maximum size permitted; Rock or gravel dumped under the track to fill in a hole

Frog - A heavy metal device used to re rail a derailed car. It could be spiked to the ties to keep it from moving. Also refers to the Heart of a switch

Fusee -A red flare used to stop a train

Hoghead - An engineer

Hoop - Three sticks fastened together in the form of a Y. With the aid of a string it was used to hand train orders and messages to trains without them having to stop.

Local - A train that stayed in one area setting out or picking up cars the area could be many miles long.

Morse - Series of dots and dashes. It could be transmitted via wire or Radio.

Plow - a snow plow. It could be a wedge or rotary type. It was shoved ahead of the engine

Puzzle switch - a switch so equipped that it could be lined to more than one Track.

Roundhouse - Large building used to service, repair and turn engines.

Siding - a track next to the main track used for the meeting and Passing of trains.

Switch - the part of a railroad track that enables trains to get from one Track to another

Torpedo - An explosive device placed on top of the rail to slow a train Down to twenty miles an hour for at least one mile

Track warrant authority - The authority for a train to move from one point to another it can contain special instructions, such as requiring it to enter a siding at the last named point.

Wait order - a train order which used to be issued to a train that instructed it to wait at one or several places until the time specified at that place. Other trains could use this information to occupy the main track, knowing the other train would wait.

Way freight - a train that stopped at many stations to set out or pick up cars